D1282973

SIX EASY PIECES
AND
SIX NOT-SO-EASY PIECES

SIX EASY PIECES
AND
SIX NOT-SO-EASY PIECES

*Essentials of Physics
Explained by Its Most
Brilliant Teacher*

RICHARD P. FEYNMAN

PERSEUS PUBLISHING
Cambridge, Massachusetts

Six Easy Pieces

A CIP catalog record for this book is available from The Library of Congress.
ISBN 0-7382-0650-4

Perseus Publishing is a member of the Perseus Books Group.
Find us on the World Wide Web at http://www.perseuspublishing.com.

10   9   8   7   6   5   4   3   2   1

Six Easy pieces
and
Six Not-So-Easy Pieces

Six Easy
PIECES

Six Easy Pieces

*Essentials of Physics
Explained by Its Most
Brilliant Teacher*

Richard P. Feynman

*Originally prepared for publication by
Robert B. Leighton and Matthew Sands*

New Introduction by Paul Davies

Contents

Publisher's Note   *vii*
Introduction by Paul Davies   *ix*
Special Preface   *xix*
Feynman's Preface   *xxv*

ONE: Atoms in Motion   1
Introduction   1
Matter is made of atoms   4
Atomic processes   10
Chemical reactions   15

TWO: Basic Physics   23
Introduction   23
Physics before 1920   27
Quantum physics   32
Nuclei and particles   38

THREE: The Relation of Physics to Other Sciences   47
Introduction   47
Chemistry   48
Biology   49
Astronomy   59
Geology   61
Psychology   63
How did it get that way?   64

Contents

FOUR: Conservation of Energy   69

What is energy?   69
Gravitational potential energy   72
Kinetic energy   80
Other forms of energy   81

FIVE: The Theory of Gravitation   89

Planetary motions   89
Kepler's laws   90
Development of dynamics   92
Newton's law of gravitation   94
Universal gravitation   98
Cavendish's experiment   104
What is gravity?   107
Gravity and relativity   112

SIX: Quantum Behavior   115

Atomic mechanics   115
An experiment with bullets   117
An experiment with waves   120
An experiment with electrons   122
The interface of electron waves   124
Watching the electrons   127
First principles of quantum mechanics   133
The uncertainty principle   136

Index   139

PUBLISHER'S NOTE

SIX EASY PIECES grew out of the need to bring to as wide an audience as possible a substantial yet non-technical physics primer based on the science of Richard Feynman. We have chosen the six easiest chapters from Feynman's celebrated landmark text, *Lectures on Physics* (originally published in 1963), which remains his most famous publication. General readers are fortunate that Feynman chose to present certain key topics in largely qualitative terms without formal mathematics, and these are brought together for SIX EASY PIECES.

Perseus Publishing would like to thank Paul Davies for his insightful introduction to this newly formed collection. Following his introduction we have chosen to reproduce two prefaces from *Lectures on Physics*, one by Feynman himself and one by two of his colleagues, because they provide context for the pieces that follow and insight into both Richard Feynman and his science.

Finally, we would like to thank the California Institute of Technology's Physics Department and Institute Archives, in particular Dr. Judith Goodstein, and Dr. Brian Hatfield, for his outstanding advice and recommendations through the development of this project.

Introduction

There is a popular misconception that science is an impersonal, dispassionate, and thoroughly objective enterprise. Whereas most other human activities are dominated by fashions, fads, and personalities, science is supposed to be constrained by agreed rules of procedure and rigorous tests. It is the results that count, not the people who produce them.

This is, of course, manifest nonsense. Science is a people-driven activity like all human endeavor, and just as subject to fashion and whim. In this case fashion is set not so much by choice of subject matter, but by the way scientists think about the world. Each age adopts its particular approach to scientific problems, usually following the trail blazed by certain dominant figures who both set the agenda and define the best methods to tackle it. Occasionally scientists attain sufficient stature that they become noticed by the general public, and when endowed with outstanding flair a scientist may become an icon for the entire scientific community. In earlier centuries Isaac Newton was an icon. Newton personified the gentleman scientist—well-connected, devoutly religious, unhurried, and methodical in his work. His style of doing science set the standard for two hundred years. In the first half of the twentieth century Albert Einstein replaced Newton as the popular scientist icon. Eccentric, dishevelled, Germanic, absent-minded, utterly absorbed in his work, and an archetypal abstract thinker, Einstein changed the way that physics is done by questioning the very concepts that define the subject.

Richard Feynman has become an icon for late twentieth-century physics—the first American to achieve this status. Born in New York in 1918 and educated on the East Coast, he was too late to participate in the Golden Age of physics, which, in the first three decades of this century, transformed our world view with the twin revolutions of the theory of relativity and quantum mechanics. These sweeping developments laid the foundations of the edifice we now call the New Physics. Feynman started with those foundations and helped build the ground floor of the New Physics. His contributions touched almost every corner of the subject and have had a deep and abiding influence over the way that physicists think about the physical universe.

Feynman was a theoretical physicist par excellence. Newton had been both experimentalist and theorist in equal measure. Einstein was quite simply contemptuous of experiment, preferring to put his faith in pure thought. Feynman was driven to develop a deep theoretical understanding of nature, but he always remained close to the real and often grubby world of experimental results. Nobody who watched the elderly Feynman elucidate the cause of the Challenger space shuttle disaster by dipping an elastic band in ice water could doubt that here was both a showman and a very practical thinker.

Initially, Feynman made a name for himself from his work on the theory of subatomic particles, specifically the topic known as quantum electrodynamics or QED. In fact, the quantum theory began with this topic. In 1900, the German physicist Max Planck proposed that light and other electromagnetic radiation, which had hitherto been regarded as waves, paradoxically behaved like tiny packets of energy, or "quanta," when interacting with matter. These particular quanta became known as photons. By the early 1930s the architects of the new quantum mechanics had worked out a mathematical scheme to describe the emission and absorption of photons by electrically charged particles such as electrons. Although this early formulation of QED enjoyed some limited success, the theory was clearly flawed. In many cases calculations gave inconsistent and even infinite answers to well-posed physical questions. It was to the

Introduction

problem of constructing a consistent theory of QED that the young Feynman turned his attention in the late 1940s.

To place QED on a sound basis it was necessary to make the theory consistent not only with the principles of quantum mechanics but with those of the special theory of relativity too. These two theories come with their own distinctive mathematical machinery, complicated systems of equations that can indeed be combined and reconciled to yield a satisfactory description of QED. Doing this was a tough undertaking, requiring a high degree of mathematical skill, and was the approach followed by Feynman's contemporaries. Feynman himself, however, took a radically different route— so radical, in fact, that he was more or less able to write down the answers straight away without using any mathematics!

To aid this extraordinary feat of intuition, Feynman invented a simple system of eponymous diagrams. Feynman diagrams are a symbolic but powerfully heuristic way of picturing what is going on when electrons, photons, and other particles interact with each other. These days Feynman diagrams are a routine aid to calculation, but in the early 1950s they marked a startling departure from the traditional way of doing theoretical physics.

The particular problem of constructing a consistent theory of quantum electrodynamics, although it was a milestone in the development of physics, was just the start. It was to define a distinctive Feynman style, a style destined to produce a string of important results from a broad range of topics in physical science. The Feynman style can best be described as a mixture of reverence and disrespect for received wisdom.

Physics is an exact science, and the existing body of knowledge, while incomplete, can't simply be shrugged aside. Feynman acquired a formidable grasp of the accepted principles of physics at a very young age, and he chose to work almost entirely on conventional problems. He was not the sort of genius to beaver away in isolation in a backwater of the discipline and to stumble across the profoundly new. His special talent was to approach essentially mainstream topics in an idiosyncratic way. This meant eschewing existing formalisms and developing his own highly intuitive approach. Whereas most theoretical physicists rely on careful mathematical calculation to provide a guide and a crutch to take them into unfamiliar territory, Feynman's attitude was almost cavalier. You get the impression that he could read nature like a book and simply report on what he found, without the tedium of complex analysis.

Indeed, in pursuing his interests in this manner Feynman displayed a healthy contempt for rigorous formalisms. It is hard to convey the depth of genius that is necessary to work like this. Theoretical physics is one of the toughest intellectual exercises, combining abstract concepts that defy visualization with extreme mathematical complexity. Only by adopting the highest standards of mental discipline can most physicists make progress. Yet Feynman appeared to ride roughshod over this strict code of practice and pluck new results like ready-made fruit from the Tree of Knowledge.

The Feynman style owed a great deal to the personality of the man. In his professional and private life he seemed to treat the world as a hugely entertaining game. The physical universe presented him with a fascinating series of puzzles and challenges, and so did his social environment. A lifelong prankster, he treated authority and the academic establishment with the same sort of disrespect he showed for stuffy mathematical formalism. Never one to suffer fools gladly, he broke the rules whenever he found them arbitrary or absurd. His autographical writings contain amusing stories of Feynman outwitting the atom-bomb security services during the war, Feynman cracking safes, Feynman disarming women with outrageously bold behavior. He treated his Nobel Prize, awarded for his work on QED, in a similar take-it-or-leave-it manner.

Alongside this distaste for formality, Feynman had a fascination with the quirky and obscure. Many will remember his obsession with the long-lost country of Tuva in Central Asia, captured so delightfully in a documentary film made near to the time of his

Introduction

death. His other passions included playing the bongo drums, painting, frequenting strip clubs, and deciphering Mayan texts.

Feynman himself did much to cultivate his distinctive persona. Although reluctant to put pen to paper, he was voluble in conversation, and loved to tell stories about his ideas and escapades. These anecdotes, accumulated over the years, helped add to his mystique and made him a proverbial legend in his own lifetime. His engaging manner endeared him greatly to students, especially the younger ones, many of whom idolized him. When Feynman died of cancer in 1988 the students at Caltech, where he had worked for most of his career, unfurled a banner with the simple message: "We love you Dick."

It was Feynman's happy-go-lucky approach to life in general and physics in particular that made him such a superb communicator. He had little time for formal lecturing or even for supervising Ph.D. students. Nevertheless he could give brilliant lectures when it suited him, deploying all the sparkling wit, penetrating insight, and irreverence that he brought to bear on his research work.

In the early 1960s Feynman was persuaded to teach an introductory physics course to Caltech freshmen and sophomores. He did so with characteristic panache and his inimitable blend of informality, zest, and off-beat humor. Fortunately, these priceless lectures were saved for posterity in book form. Though far removed in style and presentation from more conventional teaching texts, the Feynman *Lectures on Physics* were a huge success, and they excited and inspired a generation of students across the world. Three decades on, these volumes have lost nothing of their sparkle and lucidity. *Six Easy Pieces* is culled directly from *Lectures on Physics*. It is intended to give general readers a substantive taste of Feynman the Educator by drawing on the early, non-technical chapters from that landmark work. The result is a delightful volume—it serves as both a primer on physics for non-scientists and as a primer on Feynman himself.

What is most impressive about Feynman's carefully crafted exposition is the way that he can develop far-reaching physical notions from the most slender investment in concepts, and a minimum in

Introduction

the way of mathematics and technical jargon. He has the knack of finding just the right analogy or everyday illustration to bring out the essence of a deep principle, without obscuring it in incidental or irrelevant details.

The selection of topics contained in this volume is not intended as a comprehensive survey of modern physics, but as a tantalizing taste of the Feynman approach. We soon discover how he can illuminate even mundane topics like force and motion with new insights. Key concepts are illustrated by examples drawn from daily life or antiquity. Physics is continually linked to other sciences while leaving the reader in no doubt about which is the fundamental discipline.

Right at the beginning of *SIX EASY PIECES* we learn how all physics is rooted in the notion of law—the existence of an ordered universe that can be understood by the application of rational reasoning. However, the laws of physics are not transparent to us in our direct observations of nature. They are frustratingly hidden, subtly encoded in the phenomena we study. The arcane procedures of the physicist—a mixture of carefully designed experimentation and mathematical theorizing—are needed to unveil the underlying lawlike reality.

Possibly the best-known law of physics is Newton's inverse square law of gravitation, discussed here in Chapter Five. The topic is introduced in the context of the solar system and Kepler's laws of planetary motion. But gravitation is universal, applying across the cosmos, enabling Feynman to spice his account with examples from astronomy and cosmology. Commenting on a picture of a globular cluster somehow held together by unseen forces, he waxes lyrical: "If one cannot see gravitation acting here, he has no soul."

Other laws are known that refer to the various non-gravitational forces of nature that describe how particles of matter interact with each other. There is but a handful of these forces, and Feynman himself holds the considerable distinction of being one of the few scientists in history to discover a new law of physics, pertaining to the way that a weak nuclear force affects the behavior of certain subatomic particles.

Introduction

High-energy particle physics was the jewel in the crown of post-war science, at once awesome and glamorous, with its huge accelerator machines and seemingly unending list of newly discovered subatomic particles. Feynman's research was directed mostly toward making sense of the results of this enterprise. A great unifying theme among particle physicists has been the role of symmetry and conservation laws in bringing order to the subatomic zoo.

As it happens, many of the symmetries known to particle physicists were familiar already in classical physics. Chief among these are the symmetries that arise from the homogeneity of space and time. Take time: apart from cosmology, where the big bang marked the beginning of time, there is nothing in physics to distinguish one moment of time from the next. Physicists say that the world is "invariant under time translations," meaning that whether you take midnight or mid-day to be the zero of time in your measurements, it makes no difference to the description of physical phenomena. Physical processes do not depend on an absolute zero of time. It turns out that this symmetry under time translation directly implies one of the most basic, and also most useful, laws of physics: the law of conservation of energy. This law says that you can move energy around and change its form but you can't create or destroy it. Feynman makes this law crystal clear with his amusing story of Dennis the Menace who is always mischievously hiding his toy building blocks from his mother (Chapter Four).

The most challenging lecture in this volume is the last, which is an exposition on quantum physics. It is no exaggeration to say that quantum mechanics had dominated twentieth-century physics and is far and away the most successful scientific theory in existence. It is indispensable for understanding subatomic particles, atoms and nuclei, molecules and chemical bonding, the structure of solids, superconductors and superfluids, the electrical and thermal conductivity of metals and semiconductors, the structure of stars, and much else. It has practical applications ranging from the laser to the microchip. All this from a theory that at first sight—and second sight—looks absolutely crazy! Niels Bohr, one of the founders of quantum mechanics, once remarked that anybody who is not shocked by the theory hasn't understood it.

The problem is that quantum ideas strike at the very heart of what we might call common-sense reality. In particular, the idea that physical objects such as electrons or atoms enjoy an independent existence, with a complete set of physical properties at all times, is called into question. For example, an electron cannot have a position in space and a well-defined speed at the same moment. If you look for where an electron is located, you will find it at a place, and if you measure its speed you will obtain a definite answer, but you cannot make both observations at once. Nor is it meaningful to attribute definite yet unknown values for the position and speed to an electron in the absence of a complete set of observations.

This indeterminism in the very nature of atomic particles is encapsulated by Heisenberg's celebrated uncertainty principle. This puts strict limits on the precision with which properties such as position and speed can be simultaneously known. A sharp value for position smears the range of possible values of speed and vice versa. Quantum fuzziness shows up in the way electrons, photons, and other particles move. Certain experiments can reveal them taking definite paths through space, after the fashion of bullets following trajectories toward a target. But other experimental arrangements reveal that these entities can also behave like waves, showing characteristic patterns of diffraction and interference.

Feynman's masterly analysis of the famous "two-slit" experiment, which teases out the "shocking" wave-particle duality in its starkest form, has become a classic in the history of scientific exposition. With a few very simple ideas, Feynman manages to take the reader to the very heart of the quantum mystery, and leaves us dazzled by the paradoxical nature of reality that it exposes.

Although quantum mechanics had made the textbooks by the early 1930s, it is typical of Feynman that, as a young man, he preferred to refashion the theory for himself in an entirely new guise. The Feynman method has the virtue that it provides us with a vivid picture of nature's quantum trickery at work. The idea is that

Introduction

the path of a particle through space is not generally well-defined in quantum mechanics. We can imagine a freely moving electron, say, not merely traveling in a straight line between A and B as common sense would suggest, but taking a variety of wiggly routes. Feynman invites us to imagine that somehow the electron explores all possible routes, and in the absence of an observation about which path is taken we must suppose that all these alternative paths somehow contribute to the reality. So when an electron arrives at a point in space—say a target screen—many different histories must be integrated together to create this one event.

Feynman's so-called path-integral, or sum-over-histories approach to quantum mechanics, set this remarkable concept out as a mathematical procedure. It remained more or less a curiosity for many years, but as physicists pushed quantum mechanics to its limits—applying it to gravitation and even cosmology—so the Feynman approach turned out to offer the best calculational tool for describing a quantum universe. History may well judge that, among his many outstanding contributions to physics, the path-integral formulation of quantum mechanics is the most significant.

Many of the ideas discussed in this volume are deeply philosophical. Yet Feynman had an abiding suspicion of philosophers. I once had occasion to tackle him about the nature of mathematics and the laws of physics, and whether abstract mathematical laws could be considered to enjoy an independent Platonic existence. He gave a spirited and skillful description of why this indeed appears so but soon backed off when I pressed him to take a specific philosophical position. He was similarly wary when I attempted to draw him out on the subject of reductionism. With hindsight, I believe that Feynman was not, after all, contemptuous of philosophical problems. But, just as he was able to do fine mathematical physics without systematic mathematics, so he produced some fine philosophical insights without systematic philosophy. It was formalism he disliked, not content.

It is unlikely that the world will see another Richard Feynman. He was very much a man of his time. The Feynman style worked

Introduction

well for a subject that was in the process of consolidating a revolution and embarking on the far-reaching exploration of its consequences. Postwar physics was secure in its foundations; mature in its theoretical structures, yet wide open for kibitzing exploitation. Feynman entered a wonderland of abstract concepts and imprinted his personal brand of thinking upon many of them. This book provides a unique glimpse into the mind of a remarkable human being.

September 1994   PAUL DAVIES

Special Preface

(from *Lectures on Physics*)

Toward the end of his life, Richard Feynman's fame had transcended the confines of the scientific community. His exploits as a member of the commission investigating the space shuttle Challenger disaster gave him widespread exposure; similarly, a best-selling book about his picaresque adventures made him a folk hero almost of the proportions of Albert Einstein. But back in 1961, even before his Nobel Prize increased his visibility to the general public, Feynman was more than merely famous among members of the scientific community—he was legendary. Undoubtedly, the extraordinary power of his teaching helped spread and enrich the legend of Richard Feynman.

He was a truly great teacher, perhaps the greatest of his era and ours. For Feynman, the lecture hall was a theater, and the lecturer a performer, responsible for providing drama and fireworks as well as facts and figures. He would prowl about the front of a classroom, arms waving, "the impossible combination of theoretical physicist and circus barker, all body motion and sound effects," wrote *The New York Times*. Whether he addressed an audience of students, colleagues, or the general public, for those lucky enough to see Feynman lecture in person, the experience was usually unconventional and always unforgettable, like the man himself.

He was the master of high drama, adept at riveting the attention of every lecture-hall audience. Many years ago, he taught a course in Advanced Quantum Mechanics, a large class comprised of a few registered graduate students and most of the Caltech physics faculty. During one lecture, Feynman started explaining how to represent certain complicated integrals diagrammatically: time on this axis, space on that axis, wiggly line for this straight line, etc. Having described what is known to the world of physics as a Feynman diagram, he turned around to face the class, grinning wickedly. "And this is called *THE* diagram!" Feynman had reached the denouement, and the lecture hall erupted with spontaneous applause.

For many years after the lectures that make up this book were given, Feynman was an occasional guest lecturer for Caltech's freshman physics course. Naturally, his appearances had to be kept secret so there would be room left in the hall for the registered students. At one such lecture the subject was curved-space time, and Feynman was characteristically brilliant. But the unforgettable moment came at the beginning of the lecture. The supernova of 1987 has just been discovered, and Feynman was very excited about it. He said, "Tycho Brahe had his supernova, and Kepler had his. Then there weren't any for 400 years. But now I have mine." The class fell silent, and Feynman continued on. "There are 10^{11} stars in the galaxy. That used to be a *huge* number. But it's only a hundred billion. It's less than the national deficit! We used to call them astronomical numbers. Now we should call them economical numbers." The class dissolved in laughter, and Feynman, having captured his audience, went on with his lecture.

Showmanship aside, Feynman's pedagogical technique was simple. A summation of his teaching philosophy was found among his papers in the Caltech archives, in a note he had scribbled to himself while in Brazil in 1952:

"First figure out why you want the students to learn the subject and what you want them to know, and the method will result more or less by common sense."

What came to Feynman by "common sense" were often brilliant twists that perfectly captured the essence of his point. Once, during a public lecture, he was trying to explain why one must not verify an idea using the same data that suggested the idea in the first place. Seeming to wander off the subject, Feynman began talking about license plates. "You know, the most amazing thing happened to me tonight. I was coming here, on the way to the lecture, and I came in through the parking lot. And you won't believe what happened. I saw a car with the license plate ARW 357. Can you imagine? Of all the millions of license plates in the state, what was the chance that I would see that particular one tonight? Amazing!" A point that even many scientists fail to grasp was made clear through Feynman's remarkable "common sense."

In 35 years at Caltech (from 1952 to 1987), Feynman was listed as teacher of record for 34 courses. Twenty-five of them were advanced graduate courses, strictly limited to graduate students, unless undergraduates asked permission to take them (they often did, and permission was nearly always granted). The rest were mainly introductory graduate courses. Only once did Feynman teach courses purely for undergraduates, and that was the celebrated occasion in the academic years 1961 to 1962 and 1962 to 1963, with a brief reprise in 1964, when he gave the lectures that were to become *The Feynman Lectures on Physics*.

At the time there was a consensus at Caltech that freshman and sophomore students were getting turned off rather than spurred on by their two years of compulsory physics. To remedy the situation, Feynman was asked to design a series of lectures to be given to the students over the course of two years, first to freshmen, and then to the same class as sophomores. When he agreed, it was immediately decided that the lectures should be transcribed for publication. That job turned out to be far more difficult than anyone had imagined. Turning out publishable books required a tremendous amount of work on the part of his colleagues, as well as Feynman himself, who did the final editing of every chapter.

And the nuts and bolts of running a course had to be addressed. This task was greatly complicated by the fact that Feynman had only a vague outline of what he wanted to cover. This meant that no one knew what Feynman would say until he stood in front of a lecture hall filled with students and said it. The Caltech professors who assisted him would then scramble as best they could to handle mundane details, such as making up homework problems.

Why did Feynman devote more than two years to revolutionize the way beginning physics was taught? One can only speculate, but there were probably three basic reasons. One is that he loved to have an audience, and this gave him a bigger theater than he usually had in graduate courses. The second was that he genuinely cared about students, and he simply thought that teaching freshmen was an important thing to do. The third and perhaps most important reason was the sheer challenge of reformulating physics, as he understood it, so that it could be presented to young students. This was his specialty, and was the standard by which he measured whether something was really understood. Feynman was once asked by a Caltech faculty member to explain why spin 1/2 particles obey Fermi-Dirac statistics. He gauged his audience perfectly and said, "I'll prepare a freshman lecture on it." But a few days later he returned and said, "You know, I couldn't do it. I couldn't reduce it to the freshman level. That means we really don't understand it."

This specialty of reducing deep ideas to simple, understandable terms is evident throughout *The Feynman Lectures on Physics*, but nowhere more so than in his treatment of quantum mechanics. To aficionados, what he has done is clear. He has presented, to beginning students, the path integral method, the technique of his own devising that allowed him to solve some of the most profound problems in physics. His own work using path integrals, among other achievements, led to the 1965 Nobel Prize that he shared with Julian Schwinger and Sin-Itero Tomanaga.

Through the distant veil of memory, many of the students and faculty attending the lectures have said that having two years of physics with Feynman was the experience of a lifetime. But that's not how it seemed at the time. Many of the students dreaded the class, and as the course wore on, attendance by the registered students started dropping alarmingly. But at the same time, more and more faculty and graduate students started attending. The room stayed full, and Feynman may never have known he was losing some of his intended audience. But even in Feynman's view, his pedagogical endeavor did not succeed. He wrote in the 1963 preface to the *Lectures:* "I don't think I did very well by the students." Rereading the books, one sometimes seems to catch Feynman looking over his shoulder, not at his young audience, but directly at his colleagues, saying, "Look at that! Look how I finessed that point! Wasn't that clever?" But even when he thought he was explaining things lucidly to freshmen or sophomores, it was not really they who were able to benefit most from what he was doing. It was his peers—scientists, physicists, and professors—who would be the main beneficiaries of his magnificent achievement, which was nothing less than to see physics through the fresh and dynamic perspective of Richard Feynman.

Feynman was more than a great teacher. His gift was that he was an extraordinary teacher of teachers. If the purpose in giving *The Feynman Lectures on Physics* was to prepare a roomful of under-graduate students to solve examination problems in physics, he cannot be said to have succeeded particularly well. Moreover, if the intent was for the books to serve as introductory college textbooks, he cannot be said to have achieved his goal. Nevertheless, the books have been translated into ten foreign languages and are available in four bilingual editions. Feynman himself believed that his most important contribution to physics would not be QED, or the theory of superfluid helium, or polarons, or partons. His foremost contri-bution would be the three red books of *The Feynman Lectures on Physics*. That belief fully justifies this commemorative issue of these celebrated books.

DAVID L. GOODSTEIN
GERRY NEUGEBAUER
April 1989   California Institute of Technology

Feynman's Preface

(from *Lectures on Physics*)

These are the lectures in physics that I gave last year and the year before to the freshman and sophomore classes at Caltech. The lectures are, of course, not verbatim—they have been edited, sometimes extensively and sometimes less so. The lectures form only part of the complete course. The whole group of 180 students gathered in a big lecture room twice a week to hear these lectures and then they broke up into small groups of 15 to 20 students in recitation sections under the guidance of a teaching assistant. In addition, there was a laboratory session once a week.

The special problem we tried to get at with these lectures was to maintain the interest of the very enthusiastic and rather smart students coming out of the high schools and into Caltech. They have heard a lot about how interesting and exciting physics is—the theory of relativity, quantum mechanics, and other modern ideas. By the end of two years of our previous course, many would be very discouraged because there were really very few grand, new, modern ideas presented to them. They were made to study inclined planes, electrostatics, and so forth, and after two years it was quite stultifying. The problem was whether or not we could make a course which would save the more advanced and excited student by maintaining his enthusiasm.

The lectures here are not in any way meant to be a survey course, but are very serious. I thought to address them to the most intelligent in the class and to make sure, if possible, that even the most intelligent student was unable to completely encompass everything that was in the lectures—by putting in suggestions of applications of the ideas and concepts in various directions outside the main line of attack. For this reason, though, I tried very hard to make all the statements as accurate as possible, to point out in every case where the equations and ideas fitted into the body of physics, and how— when they learned more—things would be modified. I also felt that for such students it is important to indicate what it is that they should—if they are sufficiently clever—be able to understand by deduction from what has been said before, and what is being put in as something new. When new ideas came in, I would try either to deduce them if they were deducible, or to explain that it *was* a new idea which hadn't any basis in terms of things they had already learned and which was not supposed to be provable—but was just added in.

At the start of these lectures, I assumed that the students knew something when they came out of high school—such things as geometrical optics, simple chemistry ideas, and so on. I also didn't see that there was any reason to make the lectures in a definite order, in the sense that I would not be allowed to mention something until I was ready to discuss it in detail. There was a great deal of mention of things to come, without complete discussions. These more complete discussions would come later when the preparation became more advanced. Examples are the discussions of inductance, and of energy levels, which are at first brought in in a very qualitative way and are later developed more completely.

At the same time that I was aiming at the more active student, I also wanted to take care of the fellow for whom the extra fireworks and side applications are merely disquieting and who cannot be expected to learn most of the material in the lecture at all. For such students, I wanted there to be at least a central core or backbone of material which he *could* get. Even if he didn't understand everything in a lecture, I hoped he wouldn't get nervous. I didn't expect him to understand everything, but only the central and most direct features. It takes, of course, a certain intelligence on his part to see which are the central theorems and central ideas, and which are the more advanced side issues and applications which he may understand only in later years.

In giving these lectures there was one serious difficulty: in the way the course was given, there wasn't any feedback from the students to the lecturer to indicate how well the lectures were going over. This is indeed a very serious difficulty, and I don't know how good the lectures really are. The whole thing was essentially an experiment. And if I did it again I wouldn't do it the same way—I hope I *don't* have to do it again! I think, though, that things worked out—so far as the physics is concerned—quite satisfactorily in the first year.

In the second year I was not so satisfied. In the first part of the course, dealing with electricity and magnetism, I couldn't think of any really unique or different way of doing it—of any way that would be particularly more exciting than the usual way of presenting it. So I don't think I did very much in the lectures on electricity and magnetism. At the end of the second year I had originally intended to go on, after the electricity and magnetism, by giving some more lectures on the properties of materials, but mainly to take up things like fundamental modes, solutions of the diffusion equation, vibrating systems, orthogonal functions, . . . developing the first stages of what are usually called "the mathematical methods of physics." In retrospect, I think that if I were doing it again I would go back to that original idea. But since it was not planned that I would be giving these lectures again, it was suggested that it might be a good idea to try to give an introduction to the quantum mechanics—what you will find in Volume III.

It is perfectly clear that students who will major in physics can wait until their third year for quantum mechanics. On the other hand, the argument was made that many of the students in our course study physics as a background for their primary interest in other fields. And the usual way of dealing with quantum mechanics makes that subject almost unavailable for the great majority of students because they have to take so long to learn it. Yet, in its real applications—especially in its more complex applications, such as in electrical engineering and chemistry—the full machinery of the differential equation approach is not actually used. So I tried to describe the principles of quantum mechanics in a way which wouldn't require that one first know the mathematics of partial differential equations. Even for a physicist I think that is an interesting thing to try to do—to present quantum mechanics in this reverse fashion—for several reasons which may be apparent in the lectures themselves. However, I think that the experiment in the quantum mechanics part was not completely successful—in large part because I really did not have enough time at the end (I should, for instance, have had three or four more lectures in order to deal more completely with such matters as energy bands and the spatial dependence of amplitudes). Also, I had never presented the subject this way before, so the lack of feedback was particularly serious. I now believe the quantum mechanics should be given at a later time. Maybe I'll have a chance to do it again someday. Then I'll do it right.

The reason there are no lectures on how to solve problems is because there were recitation sections. Although I did put in three lectures in the first year on how to solve problems, they are not included here. Also there was a lecture on inertial guidance which certainly belongs after the lecture on rotating systems, but which was, unfortunately, omitted. The fifth and sixth lectures are actually due to Matthew Sands, as I was out of town.

The question, of course, is how well this experiment has succeeded. My own point of view—which, however, does not seem to be shared by most of the people who worked with the students—is pessimistic. I don't think I did very well by the students. When I look at the way the majority of the students handled the problems on the examinations, I think that the system is a failure. Of course, my friends point out to me that there were one or two dozen students who—very surprisingly—understood almost everything in all of the lectures, and who were quite active in working with the material and worrying about the many points in an excited and interested way. These people have now, I believe, a first-rate background in physics—and they are, after all, the ones I was trying to get at. But then, "The power of instruction is seldom of much efficacy except in those happy dispositions where it is almost superfluous." (Gibbon)

Still, I didn't want to leave any student completely behind, as perhaps I did. I think one way we could help the students more would be by putting more hard work into developing a set of problems which would elucidate some of the ideas in the lectures. Problems give a good opportunity to fill out the material of the lectures and make more realistic, more complete, and more settled in the mind the ideas that have been exposed.

I think, however, that there isn't any solution to this problem of education other than to realize that the best teaching can be done only when there is a direct individual relationship between a student and a good teacher—a situation in which the student discusses the ideas, thinks about the things, and talks about the things. It's impossible to learn very much by simply sitting in a lecture, or even by simply doing problems that are assigned. But in our modern times we have so many students to teach that we have to try to find some substitute for the ideal. Perhaps my lectures can make some contribution. Perhaps in some small place where there are individual teachers and students, they may get some inspiration or some ideas from the lectures. Perhaps they will have fun thinking them through—or going on to develop some of the ideas further.

June 1963   RICHARD P. FEYNMAN

One

ATOMS IN MOTION

Introduction

▷ This two-year course in physics is presented from the point of view that you, the reader, are going to be a physicist. This is not necessarily the case of course, but that is what every professor in every subject assumes! If you are going to be a physicist, you will have a lot to study: two hundred years of the most rapidly developing field of knowledge that there is. So much knowledge, in fact, that you might think that you cannot learn all of it in four years, and truly you cannot; you will have to go to graduate school too!

Surprisingly enough, in spite of the tremendous amount of work that has been done for all this time it is possible to condense the enormous mass of results to a large extent—that is, to find *laws* which summarize all our knowledge. Even so, the laws are so hard to grasp that it is unfair to you to start exploring this tremendous subject without some kind of map or outline of the relationship of one part of the subject of science to another. Following these preliminary remarks, the first three chapters will therefore outline the relation of physics to the rest of the sciences, the relations of the sciences to each other, and the meaning of science, to help us develop a "feel" for the subject.

You might ask why we cannot teach physics by just giving the basic laws on page one and then showing how they work in all possible circumstances, as we do in Euclidean geometry, where we

1

state the axioms and then make all sorts of deductions. (So, not satisfied to learn physics in four years, you want to learn it in four minutes?) We cannot do it in this way for two reasons. First, we do not yet *know* all the basic laws: there is an expanding frontier of ignorance. Second, the correct statement of the laws of physics involves some very unfamiliar ideas which require advanced mathematics for their description. Therefore, one needs a considerable amount of preparatory training even to learn what the *words* mean. No, it is not possible to do it that way. We can only do it piece by piece.

Each piece, or part, of the whole of nature is always merely an *approximation* to the complete truth, or the complete truth so far as we know it. In fact, everything we know is only some kind of approximation, because *we know that we do not know all the laws* as yet. Therefore, things must be learned only to be unlearned again or, more likely, to be corrected.

The principle of science, the definition, almost, is the following: *The test of all knowledge is experiment*. Experiment is the *sole judge* of scientific "truth." But what is the source of knowledge? Where do the laws that are to be tested come from? Experiment, itself, helps to produce these laws, in the sense that it gives us hints. But also needed is *imagination* to create from these hints the great generalizations—to guess at the wonderful, simple, but very strange patterns beneath them all, and then to experiment to check again whether we have made the right guess. This imagining process is so difficult that there is a division of labor in physics: there are *theoretical* physicists who imagine, deduce, and guess at new laws, but do not experiment; and then there are *experimental* physicists who experiment, imagine, deduce, and guess.

We said that the laws of nature are approximate: that we first find the "wrong" ones, and then we find the "right" ones. Now, how can an experiment be "wrong"? First, in a trivial way: if something is wrong with the apparatus that you did not notice. But these things are easily fixed, and checked back and forth. So without snatching

Atoms in Motion

at such minor things, how *can* the results of an experiment be wrong? Only by being inaccurate. For example, the mass of an object never seems to change: a spinning top has the same weight as a still one. So a "law" was invented: mass is constant, independent of speed. That "law" is now found to be incorrect. Mass is found to increase with velocity, but appreciable increases require velocities near that of light. A *true* law is: if an object moves with a speed of less than one hundred miles a second the mass is constant to within one part in a million. In some such approximate form this is a correct law. So in practice one might think that the new law makes no significant difference. Well, yes and no. For ordinary speeds we can certainly forget it and use the simple constant-mass law as a good approximation. But for high speeds we are wrong, and the higher the speed, the more wrong we are.

Finally, and most interesting, *philosophically we are completely wrong* with the approximate law. Our entire picture of the world has to be altered even though the mass changes only by a little bit. This is a very peculiar thing about the philosophy, or the ideas, behind the laws. Even a very small effect sometimes requires profound changes in our ideas.

Now, what should we teach first? Should we teach the *correct* but unfamiliar law with its strange and difficult conceptual ideas, for example the theory of relativity, four-dimensional space-time, and so on? Or should we first teach the simple "constant-mass" law, which is only approximate, but does not involve such difficult ideas? The first is more exciting, more wonderful, and more fun, but the second is easier to get at first, and is a first step to a real understanding of the second idea. This point arises again and again in teaching physics. At different times we shall have to resolve it in different ways, but at each stage it is worth learning what is now known, how accurate it is, how it fits into everything else, and how it may be changed when we learn more.

Let us now proceed with our outline, or general map, of our understanding of science today (in particular, physics, but also of other sciences on the periphery), so that when we later concentrate on some particular point we will have some idea of the background, why that particular point is interesting, and how it fits into the big structure. So, what *is* our overall picture of the world?

Matter is made of atoms

If, in some cataclysm, all of scientific knowledge were to be destroyed, and only one sentence passed on to the next generations of creatures, what statement would contain the most information in the fewest words? I believe it is the *atomic hypothesis* (or the atomic *fact*, or whatever you wish to call it) that *all things are made of atoms—little particles that move around in perpetual motion, attracting each other when they are a little distance apart, but repelling upon being squeezed into one another*. In that one sentence, you will see, there is an *enormous* amount of information about the world, if just a little imagination and thinking are applied.

To illustrate the power of the atomic idea, suppose that we have a drop of water a quarter of an inch on the side. If we look at it very closely we see nothing but water—smooth, continuous water. Even if we magnify it with the best optical microscope available— roughly two thousand times—then the water drop will be roughly forty feet across, about as big as a large room, and if we looked rather closely, we would *still* see relatively smooth water—but here and there small football-shaped things swimming back and forth. Very interesting. These are paramecia. You may stop at this point and get so curious about the paramecia with their wiggling cilia and twisting bodies that you go no further, except perhaps to magnify the paramecia still more and see inside. This, of course, is a subject for biology, but for the present we pass on and look still more closely at the water material itself, magnifying it two thousand times again. Now the drop of water extends about fifteen miles across, and if we look very closely at it we see a kind of teeming, something which no longer has a smooth appearance—it looks something like a crowd at a football game as seen from a very great distance. In

Atoms in Motion

order to see what this teeming is about, we will magnify it another two hundred and fifty times and we will see something similar to what is shown in Fig. 1–1. This is a picture of water magnified a billion times, but idealized in several ways. In the first place, the particles are drawn in a simple manner with sharp edges, which is inaccurate. Secondly, for simplicity, they are sketched almost schematically in a two-dimensional arrangement, but of course they are moving around in three dimensions. Notice that there are two kinds of "blobs" or circles to represent the atoms of oxygen (black) and hydrogen (white), and that each oxygen has two hydrogens tied to it. (Each little group of an oxygen with its two hydrogens is called a molecule.) The picture is idealized further in that the real particles in nature are continually jiggling and bouncing, turning and twisting around one another. You will have to imagine this as a dynamic rather than a static picture. Another thing that cannot be illustrated in a drawing is the fact that the particles are "stuck together"—that they attract each other, this one pulled by that one, etc. The whole group is "glued together," so to speak. On the other hand, the particles do not squeeze through each other. If you try to squeeze two of them too close together, they repel.

The atoms are 1 or 2×10^{-8} cm in radius. Now 10^{-8} cm is called an *angstrom* (just as another name), so we say they are 1 or 2 angstroms (A) in radius. Another way to remember their size is this: if an apple is magnified to the size of the earth, then the atoms in the apple are approximately the size of the original apple.

Now imagine this great drop of water with all of these jiggling particles stuck together and tagging along with each other. The water keeps its volume; it does not fall apart, because of the attraction of the molecules for each other. If the drop is on a slope, where it can move from one place to another, the water will flow, but it does not just disappear—things do not just fly apart—because of the molecular attraction. Now the jiggling motion is what we represent as *heat*: when we increase the temperature, we increase the motion. If we heat the water, the jiggling increases and the volume between the atoms increases, and if the heating continues there

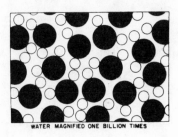

WATER MAGNIFIED ONE BILLION TIMES

Figure 1–1.

comes a time when the pull between the molecules is not enough to hold them together and they *do* fly apart and become separated from one another. Of course, this is how we manufacture steam out of water—by increasing the temperature; the particles fly apart because of the increased motion.

In Fig. 1–2 we have a picture of steam. This picture of steam fails in one respect: at ordinary atmospheric pressure there might be only a few molecules in a whole room, and there certainly would not be as many as three in this figure. Most squares this size would contain none—but we accidentally have two and a half or three in the picture (just so it would not be completely blank). Now in the case of steam we see the characteristic molecules more clearly than in the case of water. For simplicity, the molecules are drawn so that there is a 120° angle between them. In actual fact the angle is 105°3′, and the distance between the center of a hydrogen and the center of the oxygen is 0.957 A, so we know this molecule very well.

Let us see what some of the properties of steam vapor or any other gas are. The molecules, being separated from one another, will bounce against the walls. Imagine a room with a number of tennis balls (a hundred or so) bouncing around in perpetual motion. When they bombard the wall, this pushes the wall away. (Of course we would have to push the wall back.) This means that the gas exerts a jittery force which our coarse senses (not being ourselves magnified a billion times) feels only as an *average push*. In order to confine a gas we must apply a pressure. Figure 1–3 shows a

Atoms in Motion

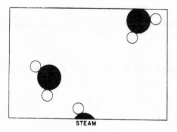

STEAM

Figure 1–2.

standard vessel for holding gases (used in all textbooks), a cylinder with a piston in it. Now, it makes no difference what the shapes of water molecules are, so for simplicity we shall draw them as tennis balls or little dots. These things are in perpetual motion in all directions. So many of them are hitting the top piston all the time that to keep it from being patiently knocked out of the tank by this continuous banging, we shall have to hold the piston down by a certain force, which we call the *pressure* (really, the pressure times the area is the force). Clearly, the force is proportional to the area, for if we increase the area but keep the number of molecules per cubic centimeter the same, we increase the number of collisions with the piston in the same proportion as the area was increased.

Now let us put twice as many molecules in this tank, so as to double the density, and let them have the same speed, i.e., the same

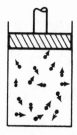

Figure 1–3.

temperature. Then, to a close approximation, the number of collisions will be doubled, and since each will be just as "energetic" as before, the pressure is proportional to the density. If we consider the true nature of the forces between the atoms, we would expect a slight decrease in pressure because of the attraction between the atoms, and a slight increase because of the finite volume they occupy. Nevertheless, to an excellent approximation, if the density is low enough that there are not many atoms, *the pressure is proportional to the density*.

We can also see something else: If we increase the temperature without changing the density of the gas, i.e., if we increase the speed of the atoms, what is going to happen to the pressure? Well, the atoms hit harder because they are moving faster, and in addition they hit more often, so the pressure increases. You see how simple the ideas of atomic theory are.

Let us consider another situation. Suppose that the piston moves inward, so that the atoms are slowly compressed into a smaller space. What happens when an atom hits the moving piston? Evidently it picks up speed from the collision. You can try it by bouncing a ping-pong ball from a forward-moving paddle, for example, and you will find that it comes off with more speed than that with which it struck. (Special example: if an atom happens to be standing still and the piston hits it, it will certainly move.) So the atoms are "hotter" when they come away from the piston than they were before they struck it. Therefore all the atoms which are in the vessel will have picked up speed. This means that *when we compress a gas slowly, the temperature of the gas increases*. So, under slow *compression*, a gas will *increase* in temperature, and under slow *expansion* it will *decrease* in temperature.

We now return to our drop of water and look in another direction. Suppose that we decrease the temperature of our drop of water. Suppose that the jiggling of the molecules of the atoms in the water is steadily decreasing. We know that there are forces of attraction between the atoms, so that after a while they will not be able to jiggle so well. What will happen at very low temperatures is

Atoms in Motion

indicated in Fig. 1–4: the molecules lock into a new pattern which is *ice*. This particular schematic diagram of ice is wrong because it is in two dimensions, but it is right qualitatively. The interesting point is that the material has a *definite place for every atom*, and you can easily appreciate that if somehow or other we were to hold all the atoms at one end of the drop in a certain arrangement, each atom in a certain place, then because of the structure of interconnections, which is rigid, the other end miles away (at our magnified scale) will have a definite location. So if we hold a needle of ice at one end, the other end resists our pushing it aside, unlike the case of water, in which the structure is broken down because of the increased jiggling so that the atoms all move around in different ways. The difference between solids and liquids is, then, that in a solid the atoms are arranged in some kind of an array, called a *crystalline array*, and they do not have a random position at long distances; the position of the atoms on one side of the crystal is determined by that of other atoms millions of atoms away on the other side of the crystal. Figure 1–4 is an invented arrangement for ice, and although it contains many of the correct features of ice, it is not the true arrangement. One of the correct features is that there is a part of the symmetry that is hexagonal. You can see that if we turn the picture around an axis by 120°, the picture returns to itself. So there is a *symmetry* in the ice which accounts for the six-sided appearance of snowflakes. Another thing we can see from Fig. 1–4 is why ice

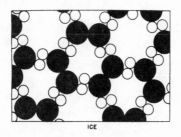

ICE

Figure 1–4.

shrinks when it melts. The particular crystal pattern of ice shown here has many "holes" in it, as does the true ice structure. When the organization breaks down, these holes can be occupied by molecules. Most simple substances, with the exception of water and type metal, *expand* upon melting, because the atoms are closely packed in the solid crystal and upon melting need more room to jiggle around, but an open structure collapses, as in the case of water.

Now although ice has a "rigid" crystalline form, its temperature can change—ice has heat. If we wish, we can change the amount of heat. What is the heat in the case of ice? The atoms are not standing still. They are jiggling and vibrating. So even though there is a definite order to the crystal—a definite structure—all of the atoms are vibrating "in place." As we increase the temperature, they vibrate with greater and greater amplitude, until they shake themselves out of place. We call this *melting*. As we decrease the temperature, the vibration decreases and decreases until, at absolute zero, there is a minimum amount of vibration that the atoms can have, but *not zero*. This minimum amount of motion that atoms can have is not enough to melt a substance, with one exception: helium. Helium merely decreases the atomic motions as much as it can, but even at absolute zero there is still enough motion to keep it from freezing. Helium, even at absolute zero, does not freeze, unless the pressure is made so great as to make the atoms squash together. If we increase the pressure, we *can* make it solidify.

Atomic processes

So much for the description of solids, liquids, and gases from the atomic point of view. However, the atomic hypothesis also describes *processes*, and so we shall now look at a number of processes from an atomic standpoint. The first process that we shall look at is associated with the surface of the water. What happens at the surface of the water? We shall now make the picture more complicated—and more realistic—by imagining that the surface is in air. Figure 1–5 shows the surface of water in air. We see the water

Atoms in Motion

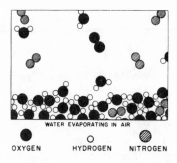

Figure 1–5.

molecules as before, forming a body of liquid water, but now we also see the surface of the water. Above the surface we find a number of things: First of all there are water molecules, as in steam. This is *water vapor*, which is always found above liquid water. (There is an equilibrium between the steam vapor and the water which will be described later.) In addition we find some other molecules—here two oxygen atoms stuck together by themselves, forming an *oxygen molecule*, there two nitrogen atoms also stuck together to make a nitrogen molecule. Air consists almost entirely of nitrogen, oxygen, some water vapor, and lesser amounts of carbon dioxide, argon, and other things. So above the water surface is the air, a gas, containing some water vapor. Now what is happening in this picture? The molecules in the water are always jiggling around. From time to time, one on the surface happens to be hit a little harder than usual, and gets knocked away. It is hard to see that happening in the picture because it is a *still* picture. But we can imagine that one molecule near the surface has just been hit and is flying out, or perhaps another one has been hit and is flying out. Thus, molecule by molecule, the water disappears—it evaporates. But if we *close* the vessel above, after a while we shall find a large number of molecules of water amongst the air molecules. From time to time, one of these vapor molecules comes flying down to the water and gets stuck again. So we see that what looks like a dead, uninteresting thing—a glass of water with a cover, that has been sitting there for perhaps twenty years—really contains a dynamic and interesting phenomenon which is going on all the time. To our eyes, our crude eyes, nothing is changing, but if we could see it a billion times magnified, we would see that from its own point of view it is always changing: molecules are leaving the surface, molecules are coming back.

Why do *we* see *no change*? Because just as many molecules are leaving as are coming back! In the long run "nothing happens." If we then take the top of the vessel off and blow the moist air away, replacing it with dry air, then the number of molecules leaving is just the same as it was before, because this depends on the jiggling of the water, but the number coming back is greatly reduced because there are so many fewer water molecules above the water. Therefore there are more going out than coming in, and the water evaporates. Hence, if you wish to evaporate water turn on the fan!

Here is something else: Which molecules leave? When a molecule leaves it is due to an accidental, extra accumulation of a little bit more than ordinary energy, which it needs if it is to break away from the attractions of its neighbors. Therefore, since those that leave have more energy than the average, the ones that are left have *less* average motion than they had before. So the liquid gradually *cools* if it evaporates. Of course, when a molecule of vapor comes from the air to the water below there is a sudden great attraction as the molecule approaches the surface. This speeds up the incoming molecule and results in generation of heat. So when they leave they take away heat; when they come back they generate heat. Of course when there is no net evaporation the result is nothing—the water is not changing temperature. If we blow on the water so as to maintain a continuous preponderance in the number evaporating, then the water is cooled. Hence, blow on soup to cool it!

Of course you should realize that the processes just described are more complicated than we have indicated. Not only does the water go into the air, but also, from time to time, one of the oxygen or nitrogen molecules will come in and "get lost" in the mass of water molecules, and work its way into the water. Thus the air dissolves in the water; oxygen and nitrogen molecules will work their way into the water and the water will contain air. If we suddenly take the air away from the vessel, then the air molecules will leave more rapidly than they come in, and in doing so will make bubbles. This is very bad for divers, as you may know.

Now we go on to another process. In Fig. 1–6 we see, from an atomic point of view, a solid dissolving in water. If we put a crystal of salt in the water, what will happen? Salt is a solid, a crystal, an organized arrangement of "salt atoms." Figure 1–7 is an illustration of the three-dimensional structure of common salt, sodium chloride. Strictly speaking, the crystal is not made of atoms, but of what we call *ions*. An ion is an atom which either has a few extra electrons or has lost a few electrons. In a salt crystal we find chlorine ions (chlorine atoms with an extra electron) and sodium ions (sodium atoms with one electron missing). The ions all stick together by electrical attraction in the solid salt, but when we put them in the water we find, because of the attractions of the negative oxygen and positive hydrogen for the ions, that some of the ions jiggle loose. In Fig. 1–6 we see a chlorine ion getting loose, and other atoms floating in the water in the form of ions. This picture was made with some care. Notice, for example, that the hydrogen ends of the water molecules are more likely to be near the chlorine ion, while near the sodium ion we are more likely to find the oxygen end, because the

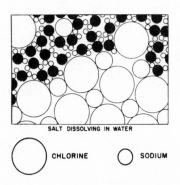

SALT DISSOLVING IN WATER

CHLORINE   SODIUM

Figure 1–6.

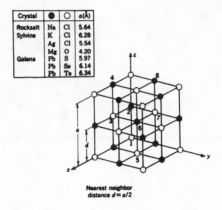

Crystal	●	○	a(Å)
Rocksalt	Na	Cl	5.64
Sylvine	K	Cl	6.28
	Ag	Cl	5.54
	Mg	O	4.20
Galena	Pb	S	5.97
	Pb	Se	6.14
	Pb	Te	6.34

Nearest neighbor
distance $d = a/2$

Figure 1–7.

sodium is positive and the oxygen end of the water is negative, and they attract electrically. Can we tell from this picture whether the salt is *dissolving in* water or *crystallizing out* of water? Of course we *cannot* tell, because while some of the atoms are leaving the crystal other atoms are rejoining it. The process is a *dynamic* one, just as in the case of evaporation, and it depends on whether there is more or less salt in the water than the amount needed for equilibrium. By equilibrium we mean that situation in which the rate at which atoms are leaving just matches the rate at which they are coming back. If there is almost no salt in the water, more atoms leave than return, and the salt dissolves. If, on the other hand, there are too many "salt atoms," more return than leave, and the salt is crystallizing.

In passing, we mention that the concept of a *molecule* of a substance is only approximate and exists only for a certain class of substances. It is clear in the case of water that the three atoms are actually stuck together. It is not so clear in the case of sodium chloride in the solid. There is just an arrangement of sodium and chlorine ions in a cubic pattern. There is no natural way to group them as "molecules of salt."

Returning to our discussion of solution and precipitation, if we increase the temperature of the salt solution, then the rate at which

Atoms in Motion

atoms are taken away is increased, and so is the rate at which atoms are brought back. It turns out to be very difficult, in general, to predict which way it is going to go, whether more or less of the solid will dissolve. Most substances dissolve more, but some substances dissolve less, as the temperature increases.

Chemical reactions

In all of the processes which have been described so far, the atoms and the ions have not changed partners, but of course there are circumstances in which the atoms do change combinations, forming new molecules. This is illustrated in Fig. 1–8. A process in which the rearrangement of the atomic partners occurs is what we call a *chemical reaction*. The other processes so far described are called physical processes, but there is no sharp distinction between the two. (Nature does not care what we call it, she just keeps on doing it.) This figure is supposed to represent carbon burning in oxygen. In the case of oxygen, *two* oxygen atoms stick together very strongly. (Why do not *three* or even *four* stick together? That is one of the very peculiar characteristics of such atomic processes. Atoms are very special: they like certain particular partners, certain particular directions, and so on. It is the job of physics to analyze why each one wants what it wants. At any rate, two oxygen atoms form, saturated and happy, a molecule.)

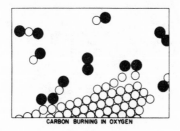

CARBON BURNING IN OXYGEN

Figure 1–8.

The carbon atoms are supposed to be in a solid crystal (which could be graphite or diamond*). Now, for example, one of the oxygen molecules can come over to the carbon, and each atom can pick up a carbon atom and go flying off in a new combination— "carbon-oxygen"—which is a molecule of the gas called carbon monoxide. It is given the chemical name CO. It is very simple: the letters "CO" are practically a picture of that molecule. But carbon attracts oxygen much more than oxygen attracts oxygen or carbon attracts carbon. Therefore in this process the oxygen may arrive with only a little energy, but the oxygen and carbon will snap together with a tremendous vengeance and commotion, and everything near them will pick up the energy. A large amount of motion energy, kinetic energy, is thus generated. This of course is *burning*; we are getting *heat* from the combination of oxygen and carbon. The heat is ordinarily in the form of the molecular motion of the hot gas, but in certain circumstances it can be so enormous that it generates *light*. That is how one gets *flames*.

In addition, the carbon monoxide is not quite satisfied. It is possible for it to attach another oxygen, so that we might have a much more complicated reaction in which the oxygen is combining with the carbon, while at the same time there happens to be a collision with a carbon monoxide molecule. One oxygen atom could attach itself to the CO and ultimately form a molecule, composed of one carbon and two oxygens, which is designated CO_2 and called carbon dioxide. If we burn the carbon with very little oxygen in a very rapid reaction (for example, in an automobile engine, where the explosion is so fast that there is not time for it to make carbon dioxide) a considerable amount of carbon monoxide is formed. In many such rearrangements, a very large amount of energy is released, forming explosions, flames, etc., depending on the reactions. Chemists have studied these arrangements of the atoms, and found that every substance is some type of *arrangement of atoms*.

---

* One *can* burn a diamond in air.

Atoms in Motion

To illustrate this idea, let us consider another example. If we go into a field of small violets, we know what "that smell" is. It is some kind of *molecule*, or arrangement of atoms, that has worked its way into our noses. First of all, *how* did it work its way in? That is rather easy. If the smell is some kind of molecule in the air, jiggling around and being knocked every which way, it might have *accidentally* worked its way into the nose. Certainly it has no particular desire to get into our nose. It is merely one helpless part of a jostling crowd of molecules, and in its aimless wanderings this particular chunk of matter happens to find itself in the nose.

Now chemists can take special molecules like the odor of violets, and analyze them and tell us the *exact arrangement* of the atoms in space. We know that the carbon dioxide molecule is straight and symmetrical: O—C—O. (That can be determined easily, too, by physical methods.) However, even for the vastly more complicated arrangements of atoms that there are in chemistry, one can, by a long, remarkable process of detective work, find the arrangements of the atoms. Figure 1–9 is a picture of the air in the neighborhood of a violet; again we find nitrogen and oxygen in the air, and water vapor. (Why is there water vapor? Because the violet is *wet*. All plants transpire.) However, we also see a "monster" composed of carbon atoms, hydrogen atoms, and oxygen atoms, which have picked a certain particular pattern in which to be arranged. It is a much more complicated arrangement than that of carbon dioxide;

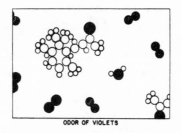

ODOR OF VIOLETS

Figure 1–9.

in fact, it is an enormously complicated arrangement. Unfortunately, we cannot picture all that is really known about it chemically, because the precise arrangement of all the atoms is actually known in three dimensions, while our picture is in only two dimensions. The six carbons which form a ring do not form a flat ring, but a kind of "puckered" ring. All of the angles and distances are known. So a chemical *formula* is merely a picture of such a molecule. When the chemist writes such a thing on the blackboard, he is trying to "draw," roughly speaking, in two dimensions. For example, we see a "ring" of six carbons, and a "chain" of carbons hanging on the end, with an oxygen second from the end, three hydrogens tied to that carbon, two carbons and three hydrogens sticking up here, etc.

How does the chemist find what the arrangement is? He mixes bottles full of stuff together, and if it turns red, it tells him that it consists of one hydrogen and two carbons tied on here; if it turns blue, on the other hand, that is not the way it is at all. This is one of the most fantastic pieces of detective work that has ever been done—organic chemistry. To discover the arrangement of the atoms in these enormously complicated arrays the chemist looks at what happens when he mixes two different substances together. The physicist could never quite believe that the chemist knew what he was talking about when he described the arrangement of the atoms. For about twenty years it has been possible, in some cases, to look at such molecules (not quite as complicated as this one, but some which contain parts of it) by a physical method, and it has been possible to locate every atom, not by looking at colors, but by *measuring where they are*. And lo and behold!, the chemists are almost always correct.

It turns out, in fact, that in the odor of violets there are three slightly different molecules, which differ only in the arrangement of the hydrogen atoms.

One problem of chemistry is to name a substance, so that we will know what it is. Find a name for this shape! Not only must the name

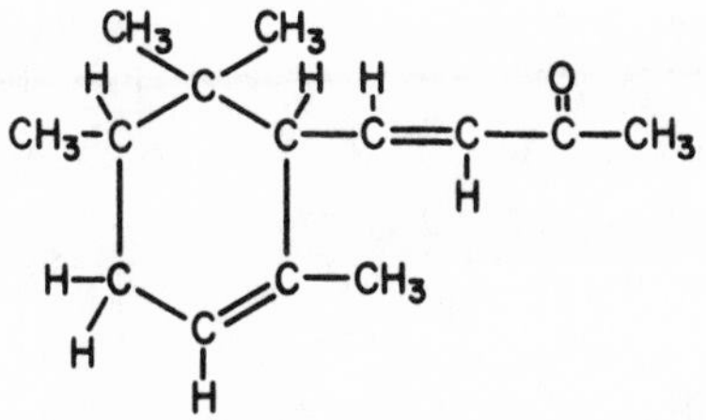

Atoms in Motion

Figure 1–10. The substance pictured is α-irone.

tell the shape, but it must also tell that here is an oxygen atom, there a hydrogen—exactly what and where each atom is. So we can appreciate that the chemical names must be complex in order to be complete. You see that the name of this thing in the more complete form that will tell you the structure of it is 4-(2, 2, 3, 6 tetramethyl-5-cyclohexanyl)-3-buten-2-one, and that tells you that this is the arrangement. We can appreciate the difficulties that the chemists have, and also appreciate the reason for such long names. It is not that they wish to be obscure, but they have an extremely difficult problem in trying to describe the molecules in words!

How do we *know* that there are atoms? By one of the tricks mentioned earlier: we make the *hypothesis* that there are atoms, and one after the other results come out the way we predict, as they ought to if things *are* made of atoms. There is also somewhat more direct evidence, a good example of which is the following: The atoms are so small that you cannot see them with a light microscope—in fact, not even with an *electron* microscope. (With a light microscope you can only see things which are much bigger.) Now if the atoms are always in motion, say in water, and we put a big ball of something in the water, a ball much bigger than the atoms, the ball will jiggle around—much as in a push ball game, where a great big ball is pushed around by a lot of people. The people are pushing in various directions, and the ball moves around the field in an irregular fashion. So, in the same way, the "large ball" will move because of the inequalities of the collisions on one side to the other, from one moment to the next. Therefore, if we look at very tiny particles (colloids) in water through an excellent microscope, we see a perpetual jiggling of the particles, which is the result of the bombardment of the atoms. This is called the *Brownian motion*.

We can see further evidence for atoms in the structure of crystals. In many cases the structures deduced by x-ray analysis agree in their spatial "shapes" with the forms actually exhibited by crystals as they occur in nature. The angles between the various "faces" of a crystal agree, within seconds of arc, with angles deduced on the assumption that a crystal is made of many "layers" of atoms.

Everything is made of atoms. That is the key hypothesis. The most important hypothesis in all of biology, for example, is that *everything that animals do, atoms do*. In other words, *there is nothing that living things do that cannot be understood from the point of view that they are made of atoms acting according to the laws of physics.* This was not known from the beginning: it took some experimenting and theorizing to suggest this hypothesis, but now it is accepted, and it is the most useful theory for producing new ideas in the field of biology.

If a piece of steel or a piece of salt, consisting of atoms one next to the other, can have such interesting properties; if water—which is nothing but these little blobs, mile upon mile of the same thing over the earth—can form waves and foam, and make rushing noises and strange patterns as it runs over cement; if all of this, all the life of a stream of water, can be nothing but a pile of atoms, *how much more is possible?* If instead of arranging the atoms in some definite pattern, again and again repeated, on and on, or even forming little lumps of complexity like the odor of violets, we make an arrangement which is *always different* from place to place, with different kinds of atoms arranged in many ways, continually changing, not repeating, how much more marvelously is it possible that this thing might behave? Is it possible that that "thing" walking back and forth in front of you, talking to you, is a great glob of these atoms in a very complex arrangement, such that the sheer complexity of it staggers the imagination as to what it can do? When we say we are a pile of atoms, we do not mean we are *merely* a pile of atoms, because a pile of atoms which is not repeated from one to the other might well have the possibilities which you see before you in the mirror.

Two

BASIC PHYSICS

Introduction

In this chapter, we shall examine the most fundamental ideas that we have about physics—the nature of things as we see them at the present time. We shall not discuss the history of how we know that all these ideas are true; you will learn these details in due time.

The things with which we concern ourselves in science appear in myriad forms, and with a multitude of attributes. For example, if we stand on the shore and look at the sea, we see the water, the waves breaking, the foam, the sloshing motion of the water, the sound, the air, the winds and the clouds, the sun and the blue sky, and light; there is sand and there are rocks of various hardness and permanence, color and texture. There are animals and seaweed, hunger and disease, and the observer on the beach; there may be even happiness and thought. Any other spot in nature has a similar variety of things and influences. It is always as complicated as that, no matter where it is. Curiosity demands that we ask questions, that we try to put things together and try to understand this multitude of aspects as perhaps resulting from the action of a relatively small number of elemental things and forces acting in an infinite variety of combinations.

For example: Is the sand other than the rocks? That is, is the sand perhaps nothing but a great number of very tiny stones? Is the moon a great rock? If we understood rocks, would we also understand the sand and the moon? Is the wind a sloshing of the air analogous to the sloshing motion of the water in the sea? What common features do different movements have? What is common to different kinds of sound? How many different colors are there? And so on. In this way we try gradually to analyze all things, to put together things which at first sight look different, with the hope that we may be able to *reduce* the number of *different* things and thereby understand them better.

A few hundred years ago, a method was devised to find partial answers to such questions. *Observation, reason,* and *experiment* make up what we call the *scientific method.* We shall have to limit ourselves to a bare description of our basic view of what is sometimes called *fundamental physics*, or fundamental ideas which have arisen from the application of the scientific method.

What do we mean by "understanding" something? We can imagine that this complicated array of moving things which constitutes "the world" is something like a great chess game being played by the gods, and we are observers of the game. We do not know what the rules of the game are; all we are allowed to do is to *watch* the playing. Of course, if we watch long enough, we may eventually catch on to a few of the rules. *The rules of the game* are what we mean by *fundamental physics*. Even if we knew every rule, however, we might not be able to understand why a particular move is made in the game, merely because it is too complicated and our minds are limited. If you play chess you must know that it is easy to learn all the rules, and yet it is often very hard to select the best move or to understand why a player moves as he does. So it is in nature, only much more so; but we may be able at least to find all the rules. Actually, we do not have all the rules now. (Every once in a while something like castling is going on that we still do not understand.) Aside from not knowing all of the rules, what we really can explain in terms of those rules is very limited, because almost all situations are so enormously complicated that we cannot follow the plays of the game using the rules, much less tell what is going to happen next. We must, therefore, limit ourselves to the more basic question

Basic Physics

of the rules of the game. If we know the rules, we consider that we "understand" the world.

How can we tell whether the rules which we "guess" at are really right if we cannot analyze the game very well? There are, roughly speaking, three ways. First, there may be situations where nature has arranged, or we arrange nature, to be simple and to have so few parts that we can predict exactly what will happen, and thus we can check how our rules work. (In one corner of the board there may be only a few chess pieces at work, and that we can figure out exactly.)

A second good way to check rules is in terms of less specific rules derived from them. For example, the rule on the move of a bishop on a chessboard is that it moves only on the diagonal. One can deduce, no matter how many moves may be made, that a certain bishop will always be on a red square. So, without being able to follow the details, we can always check our idea about the bishop's motion by finding out whether it is always on a red square. Of course it will be, for a long time, until all of a sudden we find that it is on a *black* square (what happened of course, is that in the meantime it was captured, another pawn crossed for queening, and it turned into a bishop on a black square). That is the way it is in physics. For a long time we will have a rule that works excellently in an overall way, even when we cannot follow the details, and then some time we may discover a *new rule*. From the point of view of basic physics, the most interesting phenomena are of course in the *new* places, the places where the rules do not work—not the places where they *do* work! That is the way in which we discover new rules.

The third way to tell whether our ideas are right is relatively crude but probably the most powerful of them all. That is, by rough *approximation*. While we may not be able to tell why Alekhine moves *this particular piece*, perhaps we can *roughly* understand that he is gathering his pieces around the king to protect it, more or less, since that is the sensible thing to do in the circumstances. In the same way, we can often understand nature, more or less, without being able to see what *every little piece* is doing, in terms of our understanding of the game.

At first the phenomena of nature were roughly divided into classes, like heat, electricity, mechanics, magnetism, properties of substances, chemical phenomena, light or optics, x-rays, nuclear physics, gravitation, meson phenomena, etc. However, the aim is to see *complete nature* as different aspects of *one set* of phenomena. That is the problem in basic theoretical physics, today—to *find the laws behind experiment*; to *amalgamate these classes*. Historically, we have always been able to amalgamate them, but as time goes on new things are found. We were amalgamating very well, when all of a sudden x-rays were found. Then we amalgamated some more, and mesons were found. Therefore, at any stage of the game, it always looks rather messy. A great deal is amalgamated, but there are always many wires or threads hanging out in all directions. That is the situation today, which we shall try to describe.

Some historic examples of amalgamation are the following. First, take *heat* and *mechanics*. When atoms are in motion, the more motion, the more heat the system contains, and so *heat and all temperature effects can be represented by the laws of mechanics*. Another tremendous amalgamation was the discovery of the relation between electricity, magnetism, and light, which were found to be different aspects of the same thing, which we call today the *electromagnetic field*. Another amalgamation is the unification of chemical phenomena, the various properties of various substances, and the behavior of atomic particles, which is in the *quantum mechanics of chemistry*.

The question is, of course, is it going to be possible to amalgamate *everything*, and merely discover that this world represents different aspects of *one* thing? Nobody knows. All we know is that as we go along, we find that we can amalgamate pieces, and then we find some pieces that do not fit, and we keep trying to put the jigsaw puzzle together. Whether there are a finite number of pieces, and whether there is even a border to the puzzle, is of course unknown.

Basic Physics

It will never be known until we finish the picture, if ever. What we wish to do here is to see to what extent this amalgamation process has gone on, and what the situation is at present, in understanding basic phenomena in terms of the smallest set of principles. To express it in a simple manner, *what are things made of and how few elements are there?*

Physics before 1920

It is a little difficult to begin at once with the present view, so we shall first see how things looked in about 1920 and then take a few things out of that picture. Before 1920, our world picture was something like this: The "stage" on which the universe goes is the three-dimensional *space* of geometry, as described by Euclid, and things change in a medium called *time*. The elements on the stage are *particles*, for example the atoms, which have some *properties*. First, the property of inertia: if a particle is moving it keeps on going in the same direction unless *forces* act upon it. The second element, then, is *forces*, which were then thought to be of two varieties: First, an enormously complicated, detailed kind of interaction force which held the various atoms in different combinations in a complicated way, which determined whether salt would dissolve faster or slower when we raise the temperature. The other force that was known was a long-range interaction—a smooth and quiet attraction—which varied inversely as the square of the distance, and was called *gravitation*. This law was known and was very simple. *Why* things remain in motion when they are moving, or *why* there is a law of gravitation was, of course, not known.

A description of nature is what we are concerned with here. From this point of view, then, a gas, and indeed *all* matter, is a myriad of moving particles. Thus many of the things we saw while standing at the seashore can immediately be connected. First the pressure: this comes from the collisions of the atoms with the walls or whatever; the drift of the atoms, if they are all moving in one direction on the average, is wind; the *random* internal motions are the *heat*. There are waves of excess density, where too many particles have collected, and so as they rush off they push up piles of particles farther out, and so on. This wave of excess density is *sound*. It is a tremendous achievement to be able to understand so much. Some of these things were described in the previous chapter.

What *kinds* of particles are there? There were considered to be 92 at that time: 92 different kinds of atoms were ultimately discovered. They had different names associated with their chemical properties.

The next part of the problem was, *what are the short-range forces?* Why does carbon attract one oxygen or perhaps two oxygens, but not three oxygens? What is the machinery of interaction between atoms? Is it gravitation? The answer is no. Gravity is entirely too weak. But imagine a force analogous to gravity, varying inversely with the square of the distance, but enormously more powerful and having one difference. In gravity everything attracts everything else, but now imagine that there are *two kinds* of "things," and that this new force (which is the electrical force, of course) has the property that likes *repel* but unlikes *attract*. The "thing" that carries this strong interaction is called *charge*.

Then what do we have? Suppose that we have two unlikes that attract each other, a plus and a minus, and that they stick very close together. Suppose we have another charge some distance away. Would it feel any attraction? It would feel *practically none*, because if the first two are equal in size, the attraction for the one and the repulsion for the other balance out. Therefore there is very little force at any appreciable distance. On the other hand, if we get *very close* with the extra charge, *attraction* arises, because the repulsion of likes and attraction of unlikes will tend to bring unlikes closer together and push likes farther apart. Then the repulsion will be *less* than the attraction. This is the reason why the atoms, which are constituted out of plus and minus electric charges, feel very little

Basic Physics

force when they are separated by appreciable distance (aside from gravity). When they come close together, they can "see inside" each other and rearrange their charges, with the result that they have a very strong interaction. The ultimate basis of an interaction between the atoms is *electrical*. Since this force is so enormous, all the plusses and all minuses will normally come together in as intimate a combination as they can. All things, even ourselves, are made of fine-grained, enormously strongly interacting plus and minus parts, all neatly balanced out. Once in a while, by accident, we may rub off a few minuses or a few plusses (usually it is easier to rub off minuses), and in those circumstances we find the force of electricity *unbalanced*, and we can then see the effects of these electrical attractions.

To give an idea of how much stronger electricity is than gravitation, consider two grains of sand, a millimeter across, thirty meters apart. If the force between them were not balanced, if everything attracted everything else instead of likes repelling, so that there were no cancellation, how much force would there be? There would be a force of *three million tons* between the two! You see, there is very, *very* little excess or deficit of the number of negative or positive charges necessary to produce appreciable electrical effects. This is, of course, the reason why you cannot see the difference between an electrically charged or uncharged thing—so few particles are involved that they hardly make a difference in the weight or size of an object.

With this picture the atoms were easier to understand. They were thought to have a "nucleus" at the center, which is positively electrically charged and very massive, and the nucleus is surrounded by a certain number of "electrons" which are very light and negatively charged. Now we go a little ahead in our story to remark that in the nucleus itself there were found two kinds of particles, protons and neutrons, almost of the same weight and very heavy. The protons are electrically charged and the neutrons are neutral. If we have an atom with six protons inside its nucleus, and this is surrounded by six electrons (the negative particles in the ordinary world of matter are all electrons, and these are very light compared with the protons and neutrons which make nuclei), this would be atom number six in the chemical table, and it is called carbon. Atom number eight is called oxygen, etc., because the chemical properties depend upon the electrons on the *outside*, and in fact only upon *how many* electrons there are. So the *chemical* properties of a substance depend only on a number, the number of electrons. (The whole list of elements of the chemists really could have been called 1, 2, 3, 4, 5, etc. Instead of saying "carbon," we could say "element six," meaning six electrons, but of course, when the elements were first discovered, it was not known that they could be numbered that way, and secondly, it would make everything look rather complicated. It is better to have names and symbols for these things, rather than to call everything by number.)

More was discovered about the electrical force. The natural interpretation of electrical interaction is that two objects simply attract each other: plus against minus. However, this was discovered to be an inadequate idea to represent it. A more adequate representation of the situation is to say that the existence of the positive charge, in some sense, distorts, or creates a "condition" in space, so that when we put the negative charge in, it feels a force. This potentiality for producing a force is called an *electric field*. When we put an electron in an electric field, we say it is "pulled." We then have two rules: (a) charges make a field, and (b) charges in fields have forces on them and move. The reason for this will become clear when we discuss the following phenomena: If we were to charge a body, say a comb, electrically, and then place a charged piece of paper at a distance and move the comb back and forth, the paper will respond by always pointing to the comb. If we shake it faster, it will be discovered that the paper is a little behind, *there is a delay* in the action. (At the first stage, when we move the comb rather slowly, we find a complication which is *magnetism*. Magnetic influences have to do with *charges in relative motion*, so magnetic forces and electric forces can really be attributed to one field, as two different aspects of exactly the same thing. A changing electric field cannot exist without magnetism.) If we move the charged paper farther out, the delay is greater. Then an interesting thing is observed. Although the forces between two charged objects should go inversely as the *square* of the distance, it is found, when we shake a charge, that the influence extends *very much farther out* than we would guess at first sight. That is, the effect falls off more slowly than the inverse square.

Here is an analogy: If we are in a pool of water and there is a floating cork very close by, we can move it "directly" by pushing the water with another cork. If you looked only at the two *corks*, all you would see would be that one moved immediately in response to the motion of the other—there is some kind of "*interaction*" between them. Of course, what we really do is to disturb the *water*; the *water* then disturbs the other cork. We could make up a "law" that if you pushed the water a little bit, an object close by in the water would move. If it were farther away, of course, the second cork would scarcely move, for we move the water *locally*. On the other hand, if we jiggle the cork a new phenomenon is involved, in which the motion of the water moves the water there, etc., and *waves* travel away, so that by jiggling, there is an influence *very much farther out*, an oscillatory influence, that cannot be understood from the direct interaction. Therefore the idea of direct interaction must be replaced with the existence of the water, or in the electrical case, with what we call the *electromagnetic field*.

The electromagnetic field can carry waves; some of these waves are *light*, others are used in *radio broadcasts*, but the general name is *electromagnetic waves*. These oscillatory waves can have various *frequencies*. The only thing that is really different from one wave to another is the *frequency of oscillation*. If we shake a charge back and forth more and more rapidly, and look at the effects, we get a whole series of different kinds of effects, which are all unified by specifying but one number, the number of oscillations per second. The usual "pickup" that we get from electric currents in the circuits in the walls of a building have a frequency of about one hundred

Frequency in oscillations/sec	Name	Rough behavior
10^2	Electrical disturbance	Field
$5 \times 10^5 - 10^6$	Radio broadcast	
10^8	FM — TV	
10^{10}	Radar	Waves
$5 \times 10^{14} - 10^{15}$	Light	
10^{18}	X-rays	
10^{21}	γ-rays, nuclear	
10^{24}	γ-rays, "artificial"	Particle
10^{27}	γ-rays, in cosmic rays	

Table 2–1. The Electromagnetic Spectrum.

cycles per second. If we increase the frequency to 500 or 1000 kilocycles (1 kilocycle = 1000 cycles) per second, we are "on the air," for this is the frequency range which is used for radio broadcasts. (Of course it has nothing to do with the *air!* We can have radio broadcasts without any air.) If we again increase the frequency, we come into the range that is used for FM and TV. Going still further, we use certain short waves, for example for *radar*. Still higher, and we do not need an instrument to "see" the stuff, we can see it with the human eye. In the range of frequency from 5×10^{14} to 5×10^{15} cycles per second our eyes would see the oscillation of the charged comb, if we could shake it that fast, as red, blue, or violet light, depending on the frequency. Frequencies below this range are called infrared, and above it, ultraviolet. The fact that we can see in a particular frequency range makes that part of the electromagnetic spectrum no more impressive than the other parts from a physicist's standpoint, but from a human standpoint, of course, it *is* more interesting. If we go up even higher in frequency, we get x-rays. X-rays are nothing but very high-frequency light. If we go still higher, we get gamma rays. These two terms, x-rays and gamma rays, are used almost synonymously. Usually electromagnetic rays coming from nuclei are called gamma rays, while those of high energy from atoms are called x-rays, but at the same frequency they are indistinguishable physically, no matter what their source. If we go to still higher frequencies, say to 10^{24} cycles per second, we find that we can make those waves artificially, for example with the synchrotron here at Caltech. We can find electromagnetic waves with stupendously high frequencies—with even a thousand times more rapid oscillation—in the waves found in *cosmic rays*. These waves cannot be controlled by us.

Quantum physics

Having described the idea of the electromagnetic field, and that this field can carry waves, we soon learn that these waves actually behave in a strange way which seems very unwavelike. At higher frequencies they behave much more like *particles!* It is *quantum mechanics*, discovered just after 1920, which explains this strange behavior. In the years before 1920, the picture of space as a three-dimensional space, and of time as a separate thing, was changed by Einstein, first into a combination which we call space-time, and then still further into a *curved* space-time to represent gravitation. So the "stage" is changed into space-time, and gravitation is presumably a modification of space-time. Then it was also found that the rules for the motions of particles were incorrect. The mechanical rules of "inertia" and "forces" are *wrong*—Newton's laws are *wrong*—in the world of atoms. Instead, it was discovered that things on a small scale behave *nothing like* things on a large scale. That is what makes physics difficult—and very interesting. It is hard because the way things behave on a small scale is so "unnatural"; we have no direct experience with it. Here things behave like nothing we know of, so that it is impossible to describe this behavior in any other than analytic ways. It is difficult, and takes a lot of imagination.

Quantum mechanics has many aspects. In the first place, the idea that a particle has a definite location and a definite speed is no longer allowed; that is wrong. To give an example of how wrong classical physics is, there is a rule in quantum mechanics that says that one cannot know both where something is and how fast it is moving. The uncertainty of the momentum and the uncertainty of the position are complementary, and the product of the two is constant. We can write the law like this: $\Delta x \, \Delta p \geq h/2\pi$, but we shall explain it in more detail later. This rule is the explanation of a very mysterious paradox: if the atoms are made out of plus and minus charges, why don't the minus charges simply sit on top of the plus charges (they attract each other) and get so close as to completely cancel them out? *Why are atoms so big?* Why is the nucleus at the center with the electrons around it? It was first thought that this was because the nucleus was so big; but no, the nucleus is *very small*. An atom has a diameter of about 10^{-8} cm. The nucleus has a diameter of about 10^{-13} cm. If we had an atom and wished to see the nucleus, we would have to magnify it until the whole atom was the size of a large room, and then the nucleus would be a bare speck which you could just about make out with the eye, but very nearly *all the weight* of the atom is in that infinitesimal *nucleus*. What keeps the electrons from simply falling in? This principle: If they were in the nucleus, we would know their position precisely, and the uncertainty principle would then require that they have a very *large* (but uncertain) momentum, i.e., a very large *kinetic energy*. With this energy they would break away from the nucleus. They make a compromise: they leave themselves a little room for this uncertainty and then jiggle with a certain amount of minimum motion in accordance with this rule. (Remember that when a crystal is cooled to absolute zero, we said that the atoms do not stop moving, they still jiggle. Why? If they stopped moving, we would know where they were and that they had zero motion, and that is against the uncertainty principle. We cannot know where they are and how fast they are moving, so they must be continually wiggling in there!)

Another most interesting change in the ideas and philosophy of

Basic Physics

science brought about by quantum mechanics is this: it is not possible to predict *exactly* what will happen in any circumstance. For example, it is possible to arrange an atom which is ready to emit light, and we can measure when it has emitted light by picking up a photon particle, which we shall describe shortly. We cannot, however, predict *when* it is going to emit the light or, with several atoms, *which one* is going to. You may say that this is because there are some internal "wheels" which we have not looked at closely enough. No, there *are* no internal wheels; nature, as we understand it today, behaves in such a way that it is *fundamentally impossible* to make a precise prediction of *exactly what will happen* in a given experiment. This is a horrible thing; in fact, philosophers have said before that one of the fundamental requisites of science is that whenever you set up the same conditions, the same thing must happen. This is simply *not true*, it is *not* a fundamental condition of science. The fact is that the same thing does not happen, that we can find only an average, statistically, as to what happens. Nevertheless, science has not completely collapsed. Philosophers, incidentally, say a great deal about what is *absolutely necessary* for science, and it is always, so far as one can see, rather naive, and probably wrong. For example, some philosopher or other said it is fundamental to the scientific effort that if an experiment is performed in, say, Stockholm, and then the same experiment is done in, say, Quito, the *same results* must occur. That is quite false. It is not necessary that *science* do that; it may be a *fact of experience*, but it is not necessary. For example, if one of the experiments is to look out at the sky and see the aurora borealis in Stockholm, you do not see it in Quito; that is a different phenomenon. "But," you say, "that is something that has to do with the outside; can you close yourself up in a box in Stockholm and pull down the shade and get any difference?" Surely. If we take a pendulum on a universal joint, and pull it out and let go, then the pendulum will swing almost in a plane, but not quite. Slowly the plane keeps changing in Stockholm, but not in Quito.

The blinds are down, too. The fact that this happened does not bring on the destruction of science. What *is* the fundamental hypothesis of science, the fundamental philosophy? We stated it in the first chapter: *the sole test of the validity of any idea is experiment*. If it turns out that most experiments work out the same in Quito as they do in Stockholm, then those "most experiments" will be used to formulate some general law, and those experiments which do not come out the same we will say were a result of the environment near Stockholm. We will invent some way to summarize the results of the experiment, and we do not have to be told ahead of time what this way will look like. If we are told that the same experiment will always produce the same result, that is all very well, but if when we try it, it does *not*, then it does *not*. We just have to take what we see, and then formulate all the rest of our ideas in terms of our actual experience.

Returning again to quantum mechanics and fundamental physics, we cannot go into details of the quantum-mechanical principles at this time, of course, because these are rather difficult to understand. We shall assume that they are there, and go on to describe what some of the consequences are. One of the consequences is that things which we used to consider as waves also behave like particles, and particles behave like waves; in fact everything behaves the same way. There is no distinction between a wave and a particle. So quantum mechanics *unifies* the idea of the field and its waves, and the particles, all into one. Now it is true that when the frequency is low, the field aspect of the phenomenon is more evident, or more useful as an approximate description in terms of everyday experiences. But as the frequency increases, the particle aspects of the phenomenon become more evident with the equipment with which we usually make the measurements. In fact, although we mentioned many frequencies, no phenomenon directly involving a frequency has yet been detected above approximately 10^{12} cycles per second. We only *deduce* the higher frequencies from the energy of the particles, by a rule which assumes that the particle-wave idea of quantum mechanics is valid.

Thus we have a new view of electromagnetic interaction. We have a new kind of *particle* to add to the electron, the proton, and the neutron. That new particle is called a *photon*. The new view of the interaction of electrons and protons that is electromagnetic theory, but with everything quantum-mechanically correct, is called *quantum electrodynamics*. This fundamental theory of the interaction of light and matter, or electric field and charges, is our greatest success so far in physics. In this one theory we have the basic rules for all ordinary phenomena except for gravitation and nuclear processes. For example, out of quantum electrodynamics come all known electrical, mechanical, and chemical laws: the laws for the collision of billiard balls, the motions of wires in magnetic fields, the specific heat of carbon monoxide, the color of neon signs, the density of salt, and the reactions of hydrogen and oxygen to make water are all consequences of this one law. All these details can be worked out if the situation is simple enough for us to make an approximation, which is almost never, but often we can understand more or less what is happening. At the present time no exceptions are found to the quantum-electrodynamic laws outside the nucleus, and there we do not know whether there is an exception because we simply do not know what is going on in the nucleus.

In principle, then, quantum electrodynamics is the theory of all chemistry, and of life, if life is ultimately reduced to chemistry and therefore just to physics because chemistry is already reduced (the part of physics which is involved in chemistry being already known). Furthermore, the same quantum electrodynamics, this great thing, predicts a lot of new things. In the first place, it tells the properties of very high-energy photons, gamma rays, etc. It predicted another very remarkable thing: besides the electron, there should be another particle of the same mass, but of opposite charge, called a *positron*, and these two, coming together, could annihilate each other with the emission of light or gamma rays. (After all, light and gamma rays are all the same, they are just different points on a frequency scale.) The generalization of this, that for each particle there is an antiparticle, turns out to be true. In the case of electrons, the antiparticle has another name—it is called a positron, but for most other particles, it is called anti-so-and-so, like antiproton or antineutron. In quantum electrodynamics, *two numbers* are put in and most of the other numbers in the world are supposed to come out. The two numbers that are put in are called the mass of the electron and the charge of the electron. Actually, that is not quite true, for we have a whole set of numbers for chemistry which tells how heavy the nuclei are. That leads us to the next part.

Nuclei and particles

What are the nuclei made of, and how are they held together? It is found that the nuclei are held together by enormous forces. When these are released, the energy released is tremendous compared with chemical energy, in the same ratio as the atomic bomb explosion is to a TNT explosion, because, of course, the atomic bomb has to do with changes inside the nucleus, while the explosion of TNT has to do with the changes of the electrons on the outside of the atoms. The question is, what are the forces which hold the protons and neutrons together in the nucleus? Just as the electrical interaction can be connected to a particle, a photon, Yukawa suggested that the forces between neutrons and protons also have a field of some kind, and that when this field jiggles it behaves like a particle. Thus there could be some other particles in the world besides protons and neutrons, and he was able to deduce the properties of these particles from the already known characteristics of nuclear forces. For example, he predicted they should have a mass of two or three hundred times that of an electron; and lo and behold, in cosmic rays there was discovered a particle of the right mass! But it later turned out to be the wrong particle. It was called a μ-meson, or muon.

Basic Physics

However, a little while later, in 1947 or 1948, another particle was found, the π-meson, or pion, which satisfied Yukawa's criterion. Besides the proton and the neutron, then, in order to get nuclear forces we must add the pion. Now, you say, "Oh great!, with this theory we make quantum nucleodynamics using the pions just like Yukawa wanted to do, and see if it works, and everything will be explained." Bad luck. It turns out that the calculations that are involved in this theory are so difficult that no one has ever been able to figure out what the consequences of the theory are, or to check it against experiment, and this has been going on now for almost twenty years!

So we are stuck with a theory, and we do not know whether it is right or wrong, but we do know that it is a *little* wrong, or at least incomplete. While we have been dawdling around theoretically, trying to calculate the consequences of this theory, the experimentalists have been discovering some things. For example, they had already discovered this μ-meson or muon, and we do not yet know where it fits. Also, in cosmic rays, a large number of other "extra" particles were found. It turns out that today we have approximately thirty particles, and it is very difficult to understand the relationships of all these particles, and what nature wants them for, or what the connections are from one to another. We do not today understand these various particles as different aspects of the same thing, and the fact that we have so many unconnected particles is a representation of the fact that we have so much unconnected information without a good theory. After the great successes of quantum electrodynamics, there is a certain amount of knowledge of nuclear physics which is rough knowledge, sort of half experience and half theory, assuming a type of force between protons and neutrons and seeing what will happen, but not really understanding where the force comes from. Aside from that, we have made very little progress. We have collected an enormous number of chemical elements. In the chemical case, there suddenly appeared a relationship among these elements which was unexpected, and which is embodied in the periodic table of Mendeléev. For example, sodium and potassium are about the same in their chemical properties and are found in the same column in the Mendeléev chart. We have been seeking a Mendeléev-type chart for the new particles. One such chart of the new particles was made independently by Gell-Mann in the U.S.A. and Nishijima in Japan. The basis of their classification is a new number, like the electric charge, which can be assigned to each particle, called its "strangeness," S. This number is conserved, like the electric charge, in reactions which take place by nuclear forces.

In Table 2–2 are listed all the particles. We cannot discuss them much at this stage, but the table will at least show you how much we do not know. Underneath each particle its mass is given in a certain unit, called the Mev. One Mev is equal to 1.782×10^{-27} gram. The reason this unit was chosen is historical, and we shall not go into it now. More massive particles are put higher up on the chart; we see that a neutron and a proton have almost the same mass. In vertical columns we have put the particles with the same electrical charge, all neutral objects in one column, all positively charged ones to the right of this one, and all negatively charged objects to the left.

Particles are shown with a solid line and "resonances" with a dashed one. Several particles have been omitted from the table. These include the important zero-mass, zero-charge particles, the photon and the graviton, which do not fall into the baryon-meson-lepton classification scheme, and also some of the newer resonances (K^*, φ, η). The antiparticles of the mesons are listed in the table, but the antiparticles of the leptons and baryons would have to be listed in another table which would look exactly like this one reflected on the zero-charge column. Although all of the particles except the electron, neutrino, photon, graviton, and proton are unstable, decay products have been shown only for the resonances. Strangeness assignments are not applicable for leptons, since they do not interact strongly with nuclei.

All particles which are together with the neutrons and protons are called *baryons*, and the following ones exist: There is a "lambda," with a mass of 1154 Mev, and three others, called sigmas,

Basic Physics

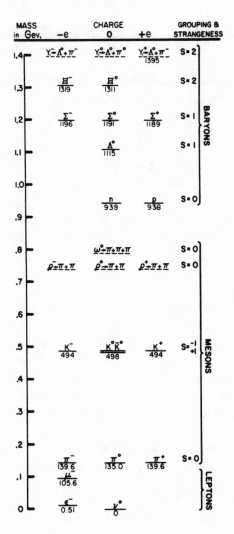

Table 2–2. Elementary Particles.

minus, neutral, and plus, with several masses almost the same. There are groups or multiplets with almost the same mass, within one or two percent. Each particle in a multiplet has the same strangeness. The first multiplet is the proton-neutron doublet, and then there is a singlet (the lambda), then the sigma triplet, and finally the xi doublet. Very recently, in 1961, even a few more particles were found. Or *are* they particles? They live so short a time, they disintegrate almost instantaneously, as soon as they are formed, that we do not know whether they should be considered as new particles, or some kind of "resonance" interaction of a certain definite energy between the Λ and π products into which they disintegrate.

In addition to the baryons the other particles which are involved in the nuclear interaction are called *mesons*. There are first the pions, which come in three varieties, positive, negative, and neutral; they form another multiplet. We have also found some new things called K-mesons, and they occur as a doublet, K^+ and K°. Also, every particle has its antiparticle, unless a particle *is its own* antiparticle. For example, the π^- and the π^+ are antiparticles, but the π° is its own antiparticle. The K^- and K^+ are antiparticles, and the K° and $\overline{K}^\circ$. In addition, in 1961 we also found some more mesons or *maybe* mesons which disintegrate almost immediately. A thing called ω which goes into three pions has a mass 780 on this scale, and somewhat less certain is an object which disintegrates into two pions. These particles, called mesons and baryons, and the antiparticles of the mesons are on the same chart, but the antiparticles of the baryons must be put on another chart, "reflected" through the charge-zero column.

Just as Mendeléev's chart was very good, except for the fact that there were a number of rare earth elements which were hanging out loose from it, so we have a number of things hanging out loose from this chart—particles which do not interact strongly in nuclei, have nothing to do with a nuclear interaction, and do not have a strong interaction (I mean the powerful kind of interaction of nuclear energy). These are called leptons, and they are the following: there

Basic Physics

is the electron, which has a very small mass on this scale, only 0.510 Mev. Then there is that other, the μ-meson, the muon, which has a mass much higher, 206 times as heavy as an electron. So far as we can tell, by all experiments so far, the difference between the electron and the muon is nothing but the mass. Everything works exactly the same for the muon as for the electron, except that one is heavier than the other. Why is there another one heavier; what is the use for it? We do not know. In addition, there is a lepton which is neutral, called a neutrino, and this particle has zero mass. In fact, it is now known that there are *two* different kinds of neutrinos, one related to electrons and the other related to muons.

Finally, we have two other particles which do not interact strongly with the nuclear ones: one is a photon, and perhaps, if the field of gravity also has a quantum-mechanical analog (a quantum theory of gravitation has not yet been worked out), then there will be a particle, a graviton, which will have zero mass.

What is this "zero mass"? The masses given here are the masses of the particles *at rest*. The fact that a particle has zero mass means, in a way, that it cannot *be* at *rest*. A photon is never at rest, it is always moving at 186,000 miles a second. We will understand more what mass means when we understand the theory of relativity, which will come in due time.

Thus we are confronted with a large number of particles, which together seem to be the fundamental constituents of matter. Fortunately, these particles are not *all* different in their *interactions* with one another. In fact, there seem to be just *four kinds* of interaction between particles which, in the order of decreasing strength, are the nuclear force, electrical interactions, the beta-decay interaction, and gravity. The photon is coupled to all charged particles and the strength of the interaction is measured by some number, which is $\frac{1}{137}$. The detailed law of this coupling is known, that is quantum electrodynamics. Gravity is coupled to all *energy*, but its coupling is extremely weak, much weaker than that of electricity. This law is also known. Then there are the so-called weak decays—beta decay, which causes the neutron to disintegrate into proton, electron, and neutrino, relatively slowly. This law is only partly known. The so-called strong interaction, the meson-baryon interaction, has a strength of 1 in this scale, and the law is completely unknown, although there are a number of known rules, such as that the number of baryons does not change in any reaction.

Coupling	Strength*	Law
Photon to charged particles	$\sim 10^{-2}$	Law known
Gravity to all energy	$\sim 10^{-40}$	Law known
Weak decays	$\sim 10^{-5}$	Law partly known
Mesons to baryons	~ 1	Law unknown (some rules known)

Table 2–3. Elementary Interactions.

This then, is the horrible condition of our physics today. To summarize it, I would say this: outside the nucleus, we seem to know all; inside it, quantum mechanics is valid—the principles of quantum mechanics have not been found to fail. The stage on which we put all of our knowledge, we would say, is relativistic space-time; perhaps gravity is involved in space-time. We do not know how the universe got started, and we have never made experiments which check our ideas of space and time accurately, below some tiny distance, so we only *know* that our ideas work above that distance. We should also add that the rules of the game are the quantum-mechanical principles, and those principles apply, so far as we can tell, to the new particles as well as to the old. The origin of the forces in nuclei leads us to new particles, but unfortunately they appear in

* The "strength" is a dimensionless measure of the coupling constant involved in each interaction ($\sim$ means "approximately").

Basic Physics

great profusion and we lack a complete understanding of their interrelationship, although we already know that there are some very surprising relationships among them. We seem gradually to be groping toward an understanding of the world of sub-atomic particles, but we really do not know how far we have yet to go in this task.

Three

THE RELATION OF PHYSICS TO OTHER SCIENCES

Introduction

Physics is the most fundamental and all-inclusive of the sciences, and has had a profound effect on all scientific development. In fact, physics is the present-day equivalent of what used to be called *natural philosophy*, from which most of our modern sciences arose. Students of many fields find themselves studying physics because of the basic role it plays in all phenomena. In this chapter we shall try to explain what the fundamental problems in the other sciences are, but of course it is impossible in so small a space really to deal with the complex, subtle, beautiful matters in these other fields. Lack of space also prevents our discussing the relation of physics to engineering, industry, society, and war, or even the most remarkable relationship between mathematics and physics. (Mathematics is not a science from our point of view, in the sense that it is not a *natural* science. The test of its validity is not experiment.) We must, incidentally, make it clear from the beginning that if a thing is not a science, it is not necessarily bad. For example, love is not a science. So, if something is said not to be a science, it does not mean that there is something wrong with it; it just means that it is not a science.

Chemistry

The science which is perhaps the most deeply affected by physics is chemistry. Historically, the early days of chemistry dealt almost entirely with what we now call inorganic chemistry, the chemistry of substances which are not associated with living things. Considerable analysis was required to discover the existence of the many elements and their relationships—how they make the various relatively simple compounds found in rocks, earth, etc. This early chemistry was very important for physics. The interaction between the two sciences was very great because the theory of atoms was substantiated to a large extent by experiments in chemistry. The theory of chemistry, i.e., of the reactions themselves, was summarized to a large extent in the periodic chart of Mendeléev, which brings out many strange relationships among the various elements, and it was the collection of rules as to which substance is combined with which, and how, that constituted inorganic chemistry. All these rules were ultimately explained in principle by quantum mechanics, so that theoretical chemistry is in fact physics. On the other hand, it must be emphasized that this explanation is *in principle*. We have already discussed the difference between knowing the rules of the game of chess, and being able to play. So it is that we may know the rules, but we cannot play very well. It turns out to be very difficult to predict precisely what will happen in a given chemical reaction; nevertheless, the deepest part of theoretical chemistry must end up in quantum mechanics.

There is also a branch of physics and chemistry which was developed by both sciences together, and which is extremely important. This is the method of statistics applied in a situation in which there are mechanical laws, which is aptly called *statistical mechanics*. In any chemical situation a large number of atoms are involved, and we have seen that the atoms are all jiggling around in a very random and complicated way. If we could analyze each collision, and be able to follow in detail the motion of each molecule, we might hope to figure out what would happen, but the many numbers needed to keep track of all these molecules exceeds so enormously the capacity of any computer, and certainly the capacity of the mind, that it was important to develop a method for dealing with such complicated situations. Statistical mechanics, then, is the science of the phenomena of heat, or thermodynamics. Inorganic chemistry is, as a science, now reduced essentially to what are called physical chemistry and quantum chemistry; physical chemistry to study the rates at which reactions occur and what is happening in detail (How do the molecules hit? Which pieces fly off first?, etc.), and quantum chemistry to help us understand what happens in terms of the physical laws.

The other branch of chemistry is *organic chemistry*, the chemistry of the substances which are associated with living things. For a time it was believed that the substances which are associated with living things were so marvelous that they could not be made by hand, from inorganic materials. This is not at all true—they are just the same as the substances made in inorganic chemistry, but more complicated arrangements of atoms are involved. Organic chemistry obviously has a very close relationship to the biology which supplies its substances, and to industry, and furthermore, much physical chemistry and quantum mechanics can be applied to organic as well as to inorganic compounds. However, the main problems of organic chemistry are not in these aspects, but rather in the analysis and synthesis of the substances which are formed in biological systems, in living things. This leads imperceptibly, in steps, toward biochemistry, and then into biology itself, or molecular biology.

Biology

Thus we come to the science of *biology*, which is the study of living things. In the early days of biology, the biologists had to deal with the purely descriptive problem of finding out *what* living things there were, and so they just had to count such things as the hairs of the limbs of fleas. After these matters were worked out with a great deal of interest, the biologists went into the *machinery* inside the living bodies, first from a gross standpoint, naturally, because it takes some effort to get into the finer details.

There was an interesting early relationship between physics and biology in which biology helped physics in the discovery of the *conservation of energy*, which was first demonstrated by Mayer in connection with the amount of heat taken in and given out by a living creature.

If we look at the processes of biology of living animals more closely, we see *many* physical phenomena: the circulation of blood, pumps, pressure, etc. There are nerves: we know what is happening when we step on a sharp stone, and that somehow or other the information goes from the leg up. It is interesting how that happens. In their study of nerves, the biologists have come to the conclusion that nerves are very fine tubes with a complex wall which is very thin; through this wall the cell pumps ions, so that there are positive ions on the outside and negative ions on the inside, like a capacitor. Now this membrane has an interesting property; if it "discharges" in one place, i.e., if some of the ions were able to move through one place, so that the electric voltage is reduced there, that electrical influence makes itself felt on the ions in the neighborhood, and it affects the membrane in such a way that it lets the ions through at neighboring points also. This in turn affects it farther along, etc., and so there is a wave of "penetrability" of the membrane which runs down the fiber when it is "excited" at one end by stepping on the sharp stone. This wave is somewhat analogous to a long sequence of vertical dominoes; if the end one is pushed over, that one pushes the next, etc. Of course this will transmit only one message unless the dominoes are set up again; and similarly in the nerve cell, there are processes which pump the ions slowly out again, to get the nerve ready for the next impulse. So it is that we know what we are doing (or at least where we are). Of course the electrical effects associated with

The Relation of Physics to Other Sciences

this nerve impulse can be picked up with electrical instruments, and because there *are* electrical effects, obviously the physics of electrical effects has had a great deal of influence on understanding the phenomenon.

The opposite effect is that, from somewhere in the brain, a message is sent out along a nerve. What happens at the end of the nerve? There the nerve branches out into fine little things, connected to a structure near a muscle, called an endplate. For reasons which are not exactly understood, when the impulse reaches the end of the nerve, little packets of a chemical called acetylcholine are shot off (five or ten molecules at a time) and they affect the muscle fiber and make it contract—how simple! What makes a muscle contract? A muscle is a very large number of fibers close together, containing two different substances, myosin and actomyosin, but the machinery by which the chemical reaction induced by acetylcholine can modify the dimensions of the molecule is not yet known. Thus the fundamental processes in the muscle that make mechanical motions are not known.

Biology is such an enormously wide field that there are hosts of other problems that we cannot mention at all—problems on how vision works (what the light does in the eye), how hearing works, etc. (The way in which *thinking* works we shall discuss later under psychology.) Now, these things concerning biology which we have just discussed are, from a biological standpoint, really not fundamental, at the bottom of life, in the sense that even if we understood them we still would not understand life itself. To illustrate: the men who study nerves feel their work is very important, because after all you cannot have animals without nerves. But you *can* have *life* without nerves. Plants have neither nerves nor muscles, but they are working, they are alive, just the same. So for the fundamental problems of biology we must look deeper; when we do, we discover that all living things have a great many characteristics in common. The most common feature is that they are made of *cells*, within each of which is complex machinery for doing things chemically. In plant cells, for example, there is machinery for picking up light and generating sucrose, which is consumed in the dark to keep the plant alive. When the plant is eaten the sucrose itself generates in the animal a series of chemical reactions very closely related to photosynthesis (and its opposite effect in the dark) in plants.

In the cells of living systems there are many elaborate chemical reactions, in which one compound is changed into another and another. To give some impression of the enormous efforts that have gone into the study of biochemistry, the chart in Fig. 3–1 summarizes our knowledge to date on just one small part of the many series of reactions which occur in cells, perhaps a percent or so of it.

Here we see a whole series of molecules which change from one to another in a sequence or cycle of rather small steps. It is called the Krebs cycle, the respiratory cycle. Each of the chemicals and each of the steps is fairly simple, in terms of what change is made in the molecule, but—and this is a centrally important discovery in biochemistry—these changes are *relatively difficult to accomplish in a laboratory*. If we have one substance and another very similar substance, the one does not just turn into the other, because the two forms are usually separated by an energy barrier or "hill." Consider this analogy: If we wanted to take an object from one place to another, at the same level but on the other side of a hill, we could push it over the top, but to do so requires the addition of some energy. Thus most chemical reactions do not occur, because there is what is called an *activation energy* in the way. In order to add an extra atom to our chemical requires that we get it *close* enough that some rearrangement can occur; then it will stick. But if we cannot give it enough energy to get it close enough, it will not go to completion, it will just go part way up the "hill" and back down again. However, if we could literally take the molecules in our hands and push and pull the atoms around in such a way as to open a hole to let the new atom in, and then let it snap back, we would have

The Relation of Physics to Other Sciences

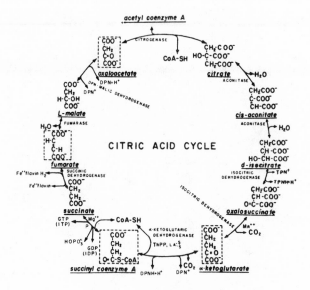

Figure 3–1. The Krebs cycle.

found another way, *around* the hill, which would not require extra energy, and the reaction would go easily. Now there actually *are*, in the cells, *very* large molecules, much larger than the ones whose changes we have been describing, which in some complicated way hold the smaller molecules just right, so that the reaction can occur easily. These very large and complicated things are called *enzymes*. (They were first called ferments, because they were originally discovered in the fermentation of sugar. In fact, some of the first reactions in the cycle were discovered there.) In the presence of an enzyme the reaction will go.

An enzyme is made of another substance called *protein*. Enzymes are very big and complicated, and each one is different, each being built to control a certain special reaction. The names of the enzymes are written in Fig. 3–1 at each reaction. (Sometimes the same enzyme may control two reactions.) We emphasize that the enzymes themselves are not involved in the reaction directly. They do not change; they merely let an atom go from one place to another. Having done so, the enzyme is ready to do it to the next molecule, like a machine in a factory. Of course, there must be a supply of certain atoms and a way of disposing of other atoms. Take hydrogen, for example: there are enzymes which have special units on them which carry the hydrogen for all chemical reactions. For example, there are three or four hydrogen-reducing enzymes which are used all over our cycle in different places. It is interesting that the machinery which liberates some hydrogen at one place will take that hydrogen and use it somewhere else.

The most important feature of the cycle of Fig. 3–1 is the transformation from GDP to GTP (guanadine-di-phosphate to guanadine-tri-phosphate) because the one substance has much more energy in it than the other. Just as there is a "box" in certain enzymes for carrying hydrogen atoms around, there are special *energy*-carrying "boxes" which involve the triphosphate group. So, GTP has more energy than GDP and if the cycle is going one way, we are producing molecules which have extra energy and which can go drive some other cycle which *requires* energy, for example the contraction of muscle. The muscle will not contract unless there is GTP. We can take muscle fiber, put it in water, and add GTP, and the fibers contract, changing GTP to GDP if the right enzymes are present. So the real system is in the GDP-GTP transformation; in the dark the GTP which has been stored up during the day is used to run the whole cycle around the other way. An enzyme, you see, does not care in which direction the reaction goes, for if it did it would violate one of the laws of physics.

Physics is of great importance in biology and other sciences for still another reason, that has to do with *experimental techniques*. In fact, if it were not for the great development of experimental physics, these biochemistry charts would not be known today. The reason is that the most useful tool of all for analyzing this fantastically complex system is to *label* the atoms which are used in the reactions. Thus, if we could introduce into the cycle some carbon dioxide which has a "green mark" on it, and then measure after three seconds where the green mark is, and again measure after ten seconds, etc., we could trace out the course of the reactions. What are the "green marks"? They are different *isotopes*. We recall that the chemical properties of atoms are determined by the number of *electrons*, not by the mass of the nucleus. But there can be, for example in carbon, six neutrons or seven neutrons, together with the six protons which all carbon nuclei have. Chemically, the two atoms C^{12} and C^{13} are the same, but they differ in weight and they have different nuclear properties, and so they are distinguishable. By using these isotopes of different weights, or even radioactive isotopes like C^{14}, which provide a more sensitive means for tracing very small quantities, it is possible to trace the reactions.

Now, we return to the description of enzymes and proteins. All proteins are not enzymes, but all enzymes are proteins. There are many proteins, such as the proteins in muscle, the structural proteins which are, for example, in cartilage and hair, skin, etc., that are not themselves enzymes. However, proteins are a very characteristic substance of life: first of all they make up all the enzymes, and second, they make up much of the rest of living material. Proteins have a very interesting and simple structure. They are a series, or chain, of different *amino acids*. There are twenty different amino acids, and they all can combine with each other to form chains in which the backbone is CO-NH, etc. Proteins are nothing but chains of various ones of these twenty amino acids. Each of the amino acids probably serves some special purpose. Some, for example, have a sulphur atom at a certain place; when two sulphur atoms are in the same protein, they form a bond, that is, they tie the chain together at two points and form a loop. Another has extra oxygen atoms which make it an acidic substance, another has a basic characteristic. Some of them have big groups hanging out to one side, so that they take up a lot of space. One of the amino acids, called prolene, is not really an amino acid, but imino acid. There is a slight difference, with the result that when prolene is in the chain, there is a kink in the chain. If we wished to manufacture a particular protein, we would give these instructions: put one of those sulphur hooks here; next, add something to take up space; then attach something to put a kink in the chain. In this way, we will get a complicated-looking chain, hooked together and having some complex structure; this is presumably just the manner in which all the various enzymes are made. One of the great triumphs in recent times (since 1960) was at last to discover the exact spatial atomic arrangement of certain proteins, which involve some fifty-six or sixty amino acids in a row. Over a thousand atoms (more nearly two thousand, if we count the hydrogen atoms) have been located in a complex pattern in two proteins. The first was hemoglobin. One of the sad aspects of this discovery is that we cannot see anything from the pattern; we do not understand why it works the way it does. Of course, that is the next problem to be attacked.

Another problem is how do the enzymes know what to be? A red-eyed fly makes a red-eyed fly baby, and so the information for the whole pattern of enzymes to make red pigment must be passed from one fly to the next. This is done by a substance in the nucleus of the cell, not a protein, called DNA (short for desoxyribose nucleic acid). This is the key substance which is passed from one cell to another (for instance, sperm cells consist mostly of DNA) and carries the information as to how to make the enzymes. DNA is the "blueprint." What does the blueprint look like and how does it work? First, the blueprint must be able to reproduce itself. Secondly, it must be able to instruct the protein. Concerning the reproduction, we might think that this proceeds like cell reproduction. Cells simply grow bigger and then divide in half. Must it be thus with DNA molecules, then, that they too grow bigger and divide in half? Every *atom* certainly does not grow bigger and divide in half! No, it is impossible to reproduce a molecule except by some more clever way.

The structure of the substance DNA was studied for a long time, first chemically to find the composition, and then with x-rays to find

The Relation of Physics to Other Sciences

the pattern in space. The result was the following remarkable discovery: The DNA molecule is a pair of chains, twisted upon each other. The backbone of each of these chains, which are analogous to the chains of proteins but chemically quite different, is a series of sugar and phosphate groups, as shown in Fig. 3–2. Now we see how the chain can contain instructions, for if we could split this chain down the middle, we would have a series *BAADC* . . . and every living thing could have a different series. Thus perhaps, in some way, the specific *instructions* for the manufacture of proteins are contained in the specific *series* of the DNA.

Attached to each sugar along the line, and linking the two chains together, are certain pairs of cross-links. However, they are not all of the same kind; there are four kinds, called adenine, thymine, cytosine, and guanine, but let us call them *A, B, C,* and *D.* The interesting thing is that only certain pairs can sit opposite each other, for example *A* with *B* and *C* with *D.* These pairs are put on the two chains in such a way that they "fit together," and have a strong energy of interaction. However, *C* will not fit with *A,* and *B* will not fit with *C;* they will only fit in pairs, *A* against *B* and *C* against *D.* Therefore if one is *C,* the other must be *D,* etc. Whatever the letters may be in one chain, each one must have its specific complementary letter on the other chain.

What then about reproduction? Suppose we split this chain in two. How can we make another one just like it? If, in the substances of the cells, there is a manufacturing department which brings up phosphate, sugar, and *A, B, C, D* units not connected in a chain, the only ones which will attach to our split chain will be the correct ones, the complements of *BAADC* . . . , namely, *ABBCD* . . . Thus what happens is that the chain splits down the middle during cell division, one half ultimately to go with one cell, the other half to end up in the other cell; when separated, a new complementary chain is made by each half-chain.

Next comes the question, precisely how does the order of the *A, B, C, D* units determine the arrangement of the amino acids in the protein? This is the central unsolved problem in biology today. The

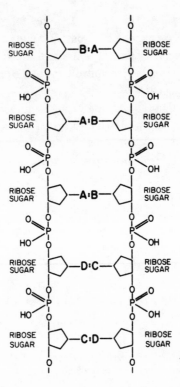

Figure 3–2. Schematic diagram of DNA.

first clues, or pieces of information, however, are these: There are in the cell tiny particles called microsomes, and it is now known that that is the place where proteins are made. But the microsomes are not in the nucleus, where the DNA and its instructions are. Something seems to be the matter. However, it is also known that little molecule pieces come off the DNA—not as long as the big DNA molecule that carries all the information itself, but like a small section of it. This is called RNA, but that is not essential. It is a kind of copy of the DNA, a short copy. The RNA, which somehow carries a message as to what kind of protein to make

The Relation of Physics to Other Sciences

goes over to the microsome; that is known. When it gets there, protein is synthesized at the microsome. That is also known. However, the details of how the amino acids come in and are arranged in accordance with a code that is on the RNA are, as yet, still unknown. We do not know how to read it. If we knew, for example, the "lineup" A, B, C, C, A, we could not tell you what protein is to be made.

Certainly no subject or field is making more progress on so many fronts at the present moment, than biology, and if we were to name the most powerful assumption of all, which leads one on and on in an attempt to understand life, it is that *all things are made of atoms*, and that everything that living things do can be understood in terms of the jigglings and wigglings of atoms.

Astronomy

In this rapid-fire explanation of the whole world, we must now turn to astronomy. Astronomy is older than physics. In fact, it got physics started by showing the beautiful simplicity of the motion of the stars and planets, the understanding of which was the *beginning* of physics. But the most remarkable discovery in all of astronomy is that *the stars are made of atoms of the same kind as those on the earth.** How was this done? Atoms liberate light which has definite

---

* How I'm rushing through this! How much each sentence in this brief story contains. "The stars are made of the same atoms as the earth." I usually pick one small topic like this to give a lecture on. Poets say science takes away from the beauty of the stars—mere globs of gas atoms. Nothing is "mere." I too can see the stars on a desert night, and feel them. But do I see less or more? The vastness of the heavens stretches my imagination—stuck on this carousel my little eye can catch one-million-year-old light. A vast pattern—of which I am a part—perhaps my stuff was belched from some forgotten star, as one is belching there. Or see them with the greater eye of Palomar, rushing all apart from some common starting point when they were perhaps all together. What is the pattern, or the meaning, or the

frequencies, something like the timbre of a musical instrument, which has definite pitches or frequencies of sound. When we are listening to several different tones we can tell them apart, but when we look with our eyes at a mixture of colors we cannot tell the parts from which it was made, because the eye is nowhere near as discerning as the ear in this connection. However, with a spectroscope we *can* analyze the frequencies of the light waves and in this way we can see the very tunes of the atoms that are in the different stars. As a matter of fact, two of the chemical elements were discovered on a star before they were discovered on the earth. Helium was discovered on the sun, whence its name, and technetium was discovered in certain cool stars. This, of course, permits us to make headway in understanding the stars, because they are made of the same kinds of atoms which are on the earth. Now we know a great deal about the atoms, especially concerning their behavior under conditions of high temperature but not very great density, so that we can analyze by statistical mechanics the behavior of the stellar substance. Even though we cannot reproduce the conditions on the earth, using the basic physical laws we often can tell precisely, or very closely, what will happen. So it is that physics aids astronomy. Strange as it may seem, we understand the distribution of matter in the interior of the sun far better than we understand the interior of the earth. What goes on *inside* a star is better understood than one might guess from the difficulty of having to look at a little dot of light through a telescope, because we can *calculate* what the atoms in the stars should do in most circumstances.

One of the most impressive discoveries was the origin of the energy of the stars, that makes them continue to burn. One of the

---

why? It does not do harm to the mystery to know a little about it. For far more marvelous is the truth than any artists of the past imagined! Why do the poets of the present not speak of it? What men are poets who can speak of Jupiter if he were like a man, but if he is an immense spinning sphere of methane and ammonia must be silent?

men who discovered this was out with his girl friend the night after he realized that *nuclear reactions* must be going on in the stars in order to make them shine. She said "Look at how pretty the stars shine!" He said, "Yes, and right now I am the only man in the world who knows *why* they shine." She merely laughed at him. She was not impressed with being out with the only man who, at that moment, knew why stars shine. Well, it is sad to be alone, but that is the way it is in this world.

It is the nuclear "burning" of hydrogen which supplies the energy of the sun; the hydrogen is converted into helium. Furthermore, ultimately, the manufacture of various chemical elements proceeds in the centers of the stars, from hydrogen. The stuff of which *we* are made, was "cooked" once, in a star, and spit out. How do we know? Because there is a clue. The proportion of the different isotopes—how much C^{12}, how much C^{13}, etc., is something which is never changed by *chemical* reactions, because the chemical reactions are so much the same for the two. The proportions are purely the result of *nuclear* reactions. By looking at the proportions of the isotopes in the cold, dead ember which we are, we can discover what the *furnace* was like in which the stuff of which we are made was formed. That furnace was like the stars, and so it is very likely that our elements were "made" in the stars and spit out in the explosions which we call novae and supernovae. Astronomy is so close to physics that we shall study many astronomical things as we go along.

Geology

We turn now to what are called *earth sciences*, or *geology*. First, meteorology and the weather. Of course the *instruments* of meteorology are physical instruments, and the development of experimental physics made these instruments possible, as was explained before. However, the theory of meteorology has never been satisfactorily worked out by the physicist. "Well," you say, "there is

nothing but air, and we know the equations of the motions of air."
Yes we do. "So if we know the condition of air today, why can't we
figure out the condition of the air tomorrow?" First, we do not
really know what the condition is today, because the air is swirling
and twisting everywhere. It turns out to be very sensitive, and even
unstable. If you have ever seen water run smoothly over a dam,
and then turn into a large number of blobs and drops as it falls,
you will understand what I mean by unstable. You know the
condition of the water before it goes over the spillway; it is
perfectly smooth; but the moment it begins to fall, where do the
drops begin? What determines how big the lumps are going to be
and where they will be? That is not known, because the water is
unstable. Even a smooth moving mass of air going over a moun-
tain turns into complex whirlpools and eddies. In many fields we
find this situation of *turbulent flow* that we cannot analyze today.
Quickly we leave the subject of weather, and discuss geology!

The question basic to geology is, what makes the earth the way
it is? The most obvious processes are in front of your very eyes, the
erosion processes of the rivers, the winds, etc. It is easy enough
to understand these, but for every bit of erosion there is an
equal amount of something else going on. Mountains are no lower
today, on the average, than they were in the past. There must
be mountain-*forming* processes. You will find, if you study geology, that there *are* mountain-forming processes and vulcanism,
which nobody understands but which is half of geology. The
phenomenon of volcanoes is really not understood. What makes
an earthquake is, ultimately, not understood. It is understood that
if something is pushing something else, it snaps and will slide—
that is all right. But what pushes, and why? The theory is that
there are currents inside the earth—circulating currents, due to
the difference in temperature inside and outside—which, in their
motion, push the surface slightly. Thus if there are two opposite
circulations next to each other, the matter will collect in the region
where they meet and make belts of mountains which are in un-

The Relation of Physics to Other Sciences

happy stressed conditions, and so produce volcanoes and earthquakes.

What about the inside of the earth? A great deal is known about the speed of earthquake waves through the earth and the density of distribution of the earth. However, physicists have been unable to get a good theory as to how dense a substance should be at the pressures that would be expected at the center of the earth. In other words, we cannot figure out the properties of matter very well in these circumstances. We do much less well with the earth than we do with the conditions of matter in the stars. The mathematics involved seems a little too difficult, so far, but perhaps it will not be too long before someone realizes that it is an important problem, and really work it out. The other aspect, of course, is that even if we did know the density, we cannot figure out the circulating currents. Nor can we really work out the properties of rocks at high pressure. We cannot tell how fast the rocks should "give"; that must all be worked out by experiment.

Psychology

Next, we consider the science of *psychology*. Incidentally, psychoanalysis is not a science: it is at best a medical process, and perhaps even more like witch-doctoring. It has a theory as to what causes disease—lots of different "spirits," etc. The witch doctor has a theory that a disease like malaria is caused by a spirit which comes into the air; it is not cured by shaking a snake over it, but quinine does help malaria. So, if you are sick, I would advise that you go to the witch doctor because he is the man in the tribe who knows the most about the disease; on the other hand, his knowledge is not science. Psychoanalysis has not been checked carefully by experiment, and there is no way to find a list of the number of cases in which it works, the number of cases in which it does not work, etc.

The other branches of psychology, which involve things like the physiology of sensation—what happens in the eye, and what happens in the brain—are, if you wish, less interesting. But some small but real progress has been made in studying them. One of the most interesting technical problems may or may not be called psychology. The central problem of the mind, if you will, or the nervous system, is this: when an animal learns something, it can do something different than it could before, and its brain cell must have changed too, if it is made out of atoms. *In what way is it different?* We do not know where to look, or what to look for, when something is memorized. We do not know what it means, or what change there is in the nervous system, when a fact is learned. This is a very important problem which has not been solved at all. Assuming, however, that there is some kind of memory thing, the brain is such an enormous mass of interconnecting wires and nerves that it probably cannot be analyzed in a straightforward manner. There is an analog of this to computing machines and computing elements, in that they also have a lot of lines, and they have some kind of element, analogous, perhaps, to the synapse, or connection of one nerve to another. This is a very interesting subject which we have not the time to discuss further—the relationship between thinking and computing machines. It must be appreciated, of course, that this subject will tell us very little about the real complexities of ordinary human behavior. All human beings are so different. It will be a long time before we get there. We must start much further back. If we could even figure out how a *dog* works, we would have gone pretty far. Dogs are easier to understand, but nobody yet knows how dogs work.

How did it get that way?

In order for physics to be useful to other sciences in a *theoretical* way, other than in the invention of instruments, the science in question must supply to the physicist a description of the object in

The Relation of Physics to Other Sciences

a physicist's language. They can say "why does a frog jump?," and the physicist cannot answer. If they tell him what a frog is, that there are so many molecules, there is a nerve here, etc., that is different. If they will tell us, more or less, what the earth or the stars are like, then we can figure it out. In order for physical theory to be of any use, we must know where the atoms are located. In order to understand the chemistry, we must know exactly what atoms are present, for otherwise we cannot analyze it. That is but one limitation, of course.

There is another *kind* of problem in the sister sciences which does not exist in physics; we might call it, for lack of a better term, the historical question. How did it get that way? If we understand all about biology, we will want to know how all the things which are on the earth got there. There is the theory of evolution, an important part of biology. In geology, we not only want to know how the mountains are forming, but how the entire earth was formed in the beginning, the origin of the solar system, etc. That, of course, leads us to want to know what kind of matter there was in the world. How did the stars evolve? What were the initial conditions? That is the problem of astronomical history. A great deal has been found out about the formation of stars, the formation of elements from which we were made, and even a little about the origin of the universe.

There is no historical question being studied in physics at the present time. We do not have a question, "Here are the laws of physics, how did they get that way?" We do not imagine, at the moment, that the laws of physics are somehow changing with time, that they were different in the past than they are at present. Of course they *may* be, and the moment we find they *are*, the historical question of physics will be wrapped up with the rest of the history of the universe, and then the physicist will be talking about the same problems as astronomers, geologists, and biologists.

Finally, there is a physical problem that is common to many fields, that is very old, and that has not been solved. It is not the problem of finding new fundamental particles, but something left over from a long time ago—over a hundred years. Nobody in physics has really been able to analyze it mathematically satisfactorily in spite of its importance to the sister sciences. It is the analysis of *circulating or turbulent fluids*. If we watch the evolution of a star, there comes a point where we can deduce that it is going to start convection, and thereafter we can no longer deduce what should happen. A few million years later the star explodes, but we cannot figure out the reason. We cannot analyze the weather. We do not know the patterns of motions that there should be inside the earth. The simplest form of the problem is to take a pipe that is very long and push water through it at high speed. We ask: to push a given amount of water through that pipe, how much pressure is needed? No one can analyze it from first principles and the properties of water. If the water flows very slowly, or if we use a thick goo like honey, then we can do it nicely. You will find that in your textbook. What we really cannot do is deal with actual, wet water running through a pipe. That is the central problem which we ought to solve some day, and we have not.

A poet once said, "The whole universe is in a glass of wine." We will probably never know in what sense he meant that, for poets do not write to be understood. But it is true that if we look at a glass of wine closely enough we see the entire universe. There are the things of physics: the twisting liquid which evaporates depending on the wind and weather, the reflections in the glass, and our imagination adds the atoms. The glass is a distillation of the earth's rocks, and in its composition we see the secrets of the universe's age, and the evolution of stars. What strange array of chemicals are in the wine? How did they come to be? There are the ferments, the enzymes, the substrates, and the products. There in wine is found the great generalization: all life is fermentation. Nobody can discover the chemistry of wine without discovering, as did Louis Pasteur, the cause of much disease. How vivid is the claret, pressing its existence into the consciousness that watches it! If our small minds, for some convenience, divide this glass of wine, this universe, into parts—physics, biology, geology, astronomy, psychology, and so on—remember that nature does not know it! So let us put it all back together, not forgetting ultimately what it is for. Let it give us one more final pleasure: drink it and forget it all!

Four

CONSERVATION OF
ENERGY

What is energy?

In this chapter, we begin our more detailed study of the different aspects of physics, having finished our description of things in general. To illustrate the ideas and the kind of reasoning that might be used in theoretical physics, we shall now examine one of the most basic laws of physics, the conservation of energy.

There is a fact, or if you wish, a *law*, governing all natural phenomena that are known to date. There is no known exception to this law—it is exact so far as we know. The law is called the *conservation of energy*. It states that there is a certain quantity, which we call energy, that does not change in the manifold changes which nature undergoes. That is a most abstract idea, because it is a mathematical principle; it says that there is a numerical quantity which does not change when something happens. It is not a description of a mechanism, or anything concrete; it is just a strange fact that we can calculate some number and when we finish watching nature go through her tricks and calculate the number again, it is the same. (Something like the bishop on a red square, and after a number of moves—details unknown—it is still on some red square.

SIX EASY PIECES

It is a law of this nature.) Since it is an abstract idea, we shall illustrate the meaning of it by an analogy.

Imagine a child, perhaps "Dennis the Menace," who has blocks which are absolutely indestructible, and cannot be divided into pieces. Each is the same as the other. Let us suppose that he has 28 blocks. His mother puts him with his 28 blocks into a room at the beginning of the day. At the end of the day, being curious, she counts the blocks very carefully, and discovers a phenomenal law—no matter what he does with the blocks, there are always 28 remaining! This continues for a number of days, until one day there are only 27 blocks, but a little investigating shows that there is one under the rug—she must look everywhere to be sure that the number of blocks has not changed. One day, however, the number appears to change—there are only 26 blocks. Careful investigation indicates that the window was open, and upon looking outside, the other two blocks are found. Another day, careful count indicates that there are 30 blocks! This causes considerable consternation, until it is realized that Bruce came to visit, bringing his blocks with him, and he left a few at Dennis' house. After she has disposed of the extra blocks, she closes the window, does not let Bruce in, and then everything is going along all right, until one time she counts and finds only 25 blocks. However, there is a box in the room, a toy box, and the mother goes to open the toy box, but the boy says "No, do not open my toy box," and screams. Mother is not allowed to open the toy box. Being extremely curious, and somewhat ingenious, she invents a scheme! She knows that a block weighs three ounces, so she weighs the box at a time when she sees 28 blocks, and it weighs 16 ounces. The next time she wishes to check, she weighs the box again, subtracts sixteen ounces and divides by three. She discovers the following:

$$\begin{pmatrix} \text{number of} \\ \text{blocks seen} \end{pmatrix} + \frac{(\text{weight of box}) - 16 \text{ ounces}}{3 \text{ ounces}} = \text{constant.} \quad (4.1)$$

Conservation of Energy

There then appear to be some new deviations, but careful study indicates that the dirty water in the bathtub is changing its level. The child is throwing blocks into the water, and she cannot see them because it is so dirty, but she can find out how many blocks are in the water by adding another term to her formula. Since the original height of the water was 6 inches and each block raises the water a quarter of an inch, this new formula would be:

$$\begin{pmatrix} \text{number of} \\ \text{blocks seen} \end{pmatrix} + \frac{(\text{weight of box}) - 16 \text{ ounces}}{3 \text{ ounces}}$$
$$+ \frac{(\text{height of water}) - 6 \text{ inches}}{1/4 \text{ inch}} = \text{constant. (4.2)}$$

In the gradual increase in the complexity of her world, she finds a whole series of terms representing ways of calculating how many blocks are in places where she is not allowed to look. As a result, she finds a complex formula, a quantity which *has to be computed*, which always stays the same in her situation.

What is the analogy of this to the conservation of energy? The most remarkable aspect that must be abstracted from this picture is that *there are no blocks*. Take away the first terms in (4.1) and (4.2) and we find ourselves calculating more or less abstract things. The analogy has the following points. First, when we are calculating the energy, sometimes some of it leaves the system and goes away, or sometimes some comes in. In order to verify the conservation of energy, we must be careful that we have not put any in or taken any out. Second, the energy has a large number of *different forms*, and there is a formula for each one. These are: gravitational energy, kinetic energy, heat energy, elastic energy, electrical energy, chemical energy, radiant energy, nuclear energy, mass energy. If we total up the formulas for each of these contributions, it will not change except for energy going in and out.

It is important to realize that in physics today, we have no knowledge of what energy *is*. We do not have a picture that energy comes in little blobs of a definite amount. It is not that way. However, there are formulas for calculating some numerical quantity, and when we add it all together it gives "28"—always the same number. It is an abstract thing in that it does not tell us the mechanism or the *reasons* for the various formulas.

Gravitational potential energy

Conservation of energy can be understood only if we have the formula for all of its forms. I wish to discuss the formula for gravitational energy near the surface of the Earth, and I wish to derive this formula in a way which has nothing to do with history but is simply a line of reasoning invented for this particular lecture to give you an illustration of the remarkable fact that a great deal about nature can be extracted from a few facts and close reasoning. It is an illustration of the kind of work theoretical physicists become involved in. It is patterned after a most excellent argument by Mr. Carnot on the efficiency of steam engines.*

Consider weight-lifting machines—machines which have the property that they lift one weight by lowering another. Let us also make a hypothesis: that *there is no such thing as perpetual motion* with these weight-lifting machines. (In fact, that there is no perpetual motion at all is a general statement of the law of conservation of energy.) We must be careful to define perpetual motion. First, let us do it for weight-lifting machines. If, when we have lifted and lowered a lot of weights and restored the machine to the original condition, we find that the net result is to have *lifted a weight*, then we have a perpetual motion machine because we can use that lifted weight to run something else. That is, *provided* the machine which lifted the weight is brought back to its exact *original condition*, and furthermore that it is completely *self-contained*—that it has not received the energy to lift that weight from some external source—like Bruce's blocks.

A very simple weight-lifting machine is shown in Fig. 4–1. This

* Our point here is not so much the result, (4.3), which in fact you may already know, as the possibility of arriving at it by theoretical reasoning.

Conservation of Energy

Figure 4–1. Simple weight-lifting machine.

machine lifts weights three units "strong." We place three units on one balance pan, and one unit on the other. However, in order to get it actually to work, we must lift a little weight off the left pan. On the other hand, we could lift a one-unit weight by lowering the three-unit weight, if we cheat a little by lifting a little weight off the other pan. Of course, we realize that with any *actual* lifting machine, we must add a little extra to get it to run. This we disregard, *temporarily*. Ideal machines, although they do not exist, do not require anything extra. A machine that we actually use can be, in a sense, *almost* reversible: that is, if it will lift the weight of three by lowering a weight of one, then it will also lift nearly the weight of one the same amount by lowering the weight of three.

We imagine that there are two classes of machines, those that are *not* reversible, which includes all real machines, and those that *are* reversible, which of course are actually not attainable no matter how careful we may be in our design of bearings, levers, etc. We suppose, however, that there is such a thing—a reversible machine—which lowers one unit of weight (a pound or any other unit) by one unit of distance, and at the same time lifts a three-unit weight. Call this reversible machine, Machine A. Suppose this particular reversible machine lifts the three-unit weight a distance X. Then suppose we have another machine, Machine B, which is not necessarily reversible, which also lowers a unit weight a unit distance, but which lifts three units a distance Y. We can now prove that Y is not higher than X; that is, it is impossible to build a machine that will lift a weight *any higher* than it will be lifted by a reversible machine. Let us see why. Let us suppose that Y were higher than X. We take a one-unit weight and lower it one unit height with Machine B, and that lifts the three-unit weight up a distance Y. Then we could lower the weight from Y to X, *obtaining free power*, and use the reversible Machine A, running backwards, to lower the three-unit weight a distance X and lift the one-unit weight by one unit height. This will put the one-unit weight back where it was before, and leave both machines ready to be used again! We would therefore have perpetual motion if Y were higher than X, which we assumed was impossible. With those assumptions, we thus deduce that Y *is not higher than* X, so that of all machines that can be designed, the reversible machine is the best.

We can also see that all reversible machines must lift to *exactly the same height*. Suppose that B were really reversible also. The argument that Y is not higher than X is, of course, just as good as it was before, but we can also make our argument the other way around, using the machines in the opposite order, and prove that X *is not higher than* Y. This, then, is a very remarkable observation because it permits us to analyze the height to which different machines are going to lift something *without looking at the interior mechanism*. We know at once that if somebody makes an enormously elaborate series of levers that lift three units a certain distance by lowering one unit by one unit distance, and we compare it with a simple lever which does the same thing and is fundamentally reversible, his machine will lift it no higher, but perhaps less high. If his machine is reversible, we also know exactly *how* high it will lift. To summarize: every reversible machine, no matter how it operates, which drops one pound one foot and lifts a three-pound weight always lifts it the same distance, X. This is clearly a universal law of great utility. The next question is, of course, what is X?

Suppose we have a reversible machine which is going to lift this distance X, three for one. We set up three balls in a rack which does not move, as shown in Fig. 4–2. One ball is held on a stage at a distance one foot above the ground. The machine can lift three balls, lowering one by a distance 1. Now, we have arranged that the platform which holds three balls has a floor and two shelves, exactly spaced at distance X, and further, that the rack which holds the balls is spaced at distance X, (a). First we roll the balls horizontally from

Conservation of Energy

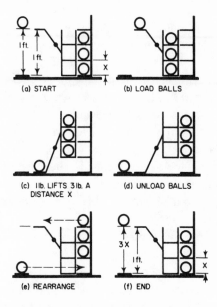

Figure 4–2. A reversible machine.

the rack to the shelves, (b), and we suppose that this takes no energy because we do not change the height. The reversible machine then operates: it lowers the single ball to the floor, and it lifts the rack a distance X, (c). Now we have ingeniously arranged the rack so that these balls are again even with the platforms. Thus we unload the balls onto the rack, (d); having unloaded the balls, we can restore the machine to its original condition. Now we have three balls on the upper three shelves and one at the bottom. But the strange thing is that, in a certain way of speaking, we have not lifted *two* of them at all because, after all, there were balls on shelves 2 and 3 before. The resulting effect has been to lift *one ball* a distance $3X$. Now, if $3X$ exceeds one foot, then we can *lower* the ball to return the machine to the initial condition, (f), and we can run the apparatus again. Therefore $3X$ cannot exceed one foot, for if $3X$ exceeds one foot we can make perpetual motion. Likewise, we can prove that *one foot cannot exceed* $3X$, by making the whole machine run the opposite way, since it is a reversible machine. Therefore $3X$ is neither *greater nor less than a foot*, and we discover then, by argument alone, the law that $X = \frac{1}{3}$ foot. The generalization is clear: one pound falls a certain distance in operating a reversible machine; then the machine can lift p pounds this distance divided by p. Another way of putting the result is that three pounds times the height lifted, which in our problem was X, is equal to one pound times the distance lowered, which is one foot in this case. If we take all the weights and multiply them by the heights at which they are now, above the floor, let the machine operate, and then multiply all the weights by all the heights again, *there will be no change*. (We have to generalize the example where we moved only one weight to the case where when we lower one we lift several different ones—but that is easy.)

We call the sum of the weights times the heights *gravitational potential energy*—the energy which an object has because of its relationship in space, relative to the earth. The formula for gravitational energy, then, so long as we are not too far from the earth (the force weakens as we go higher) is

$$\begin{pmatrix} \text{gravitational} \\ \text{potential energy} \\ \text{for one object} \end{pmatrix} = (\text{weight}) \times (\text{height}). \qquad (4.3)$$

It is a very beautiful line of reasoning. The only problem is that perhaps it is not true. (After all, nature does not *have* to go along with our reasoning.) For example, perhaps perpetual motion is, in fact, possible. Some of the assumptions may be wrong, or we may have made a mistake in reasoning, so it is always necessary to check. *It turns out experimentally*, in fact, to be true.

The general name of energy which has to do with location relative to something else is called *potential* energy. In this particular case, of course, we call it *gravitational potential energy*. If it is a question of electrical forces against which we are working, instead of gravitational forces, if we are "lifting" charges away from other charges with a lot of levers, then the energy content is called *electrical*

Conservation of Energy

potential energy. The general principle is that the change in the energy is the force times the distance that the force is pushed, and that this is a change in energy in general:

$$\begin{pmatrix}\text{change in}\\\text{energy}\end{pmatrix} = (\text{force}) \times \begin{pmatrix}\text{distance force}\\\text{acts through}\end{pmatrix} \qquad (4.4)$$

We will return to many of these other kinds of energy as we continue the course.

The principle of the conservation of energy is very useful for deducing what will happen in a number of circumstances. In high school we learned a lot of laws about pulleys and levers used in different ways. We can now see that these "laws" are *all the same thing*, and that we did not have to memorize 75 rules to figure it out. A simple example is a smooth inclined plane which is, happily, a three-four-five triangle (Fig. 4–3). We hang a one-pound weight on the inclined plane with a pulley, and on the other side of the pulley, a weight W. We want to know how heavy W must be to balance the one pound on the plane. How can we figure that out? If we say it is just balanced, it is reversible and so can move up and down, and we can consider the following situation. In the initial circumstance, (a), the one pound weight is at the bottom and weight W is at the top. When W has slipped down in a reversible way, we have a one-pound weight at the top and the weight W the slant distance, (b), or five feet, from the plane in which it was before. We *lifted* the one-pound weight only *three* feet and we lowered W pounds by *five* feet.

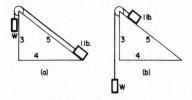

Figure 4–3. Inclined plane.

SIX EASY PIECES

Therefore $W = \frac{3}{5}$ of a pound. Note that we deduced this from the *conservation of energy*, and not from force components. Cleverness, however, is relative. It can be deduced in a way which is even more brilliant, discovered by Stevinus and inscribed on his tombstone. Figure 4–4 explains that it has to be $\frac{3}{5}$ of a pound, because the chain does not go around. It is evident that the lower part of the chain is balanced by itself, so that the pull of the five weights on one side must balance the pull of three weights on the other, or whatever the ratio of the legs. You see, by looking at this diagram, that W must be $\frac{3}{5}$ of a pound. (If you get an epitaph like that on your gravestone, you are doing fine.)

Let us now illustrate the energy principle with a more complicated problem, the screw jack shown in Fig. 4–5. A handle 20 inches long is used to turn the screw, which has 10 threads to the inch. We would like to know how much force would be needed at the handle to lift one ton (2000 pounds). If we want to lift the ton one inch, say, then we must turn the handle around ten times. When it goes around once it goes approximately 126 inches. The handle must thus travel 1260 inches, and if we used various pulleys, etc., we would be lifting our one ton with an unknown smaller weight W applied to the end of the handle. So we find out that W is about 1.6 pounds. This is a result of the conservation of energy.

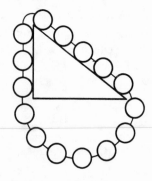

Figure 4–4. The epitaph of Stevinus.

Conservation of Energy

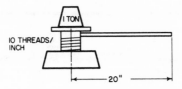

Figure 4–5. A screw jack.

Take now the somewhat more complicated example shown in Fig. 4–6. A rod or bar, 8 feet long, is supported at one end. In the middle of the bar is a weight of 60 pounds, and at a distance of two feet from the support there is a weight of 100 pounds. How hard do we have to lift the end of the bar in order to keep it balanced, disregarding the weight of the bar? Suppose we put a pulley at one end and hang a weight on the pulley. How big would the weight W have to be in order for it to balance? We imagine that the weight falls any arbitrary distance—to make it easy for ourselves suppose it goes down 4 inches—how high would the two load weights rise? The center rises 2 inches, and the point a quarter of the way from the fixed end lifts 1 inch. Therefore, the principle that the sum of the heights times the weights does not change tells us that the weight W times 4 inches down, plus 60 pounds times 2 inches up, plus 100 pounds times 1 inch has to add up to nothing:

$$- 4W + (2)(60) + (1)(100) = 0, \qquad W = 55 \text{ lb.} \qquad (4.5)$$

Thus we must have a 55-pound weight to balance the bar. In this way we can work out the laws of "balance"—the statics of complicated bridge arrangements, and so on. This approach is called the *principle of virtual work*, because in order to apply this argument we had to *imagine* that the structure moves a little—even though it is

SIX EASY PIECES

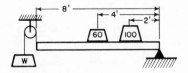

Figure 4–6. Weighted rod supported on one end.

not *really* moving or even *movable*. We use the very small imagined motion to apply the principle of conservation of energy.

Kinetic energy

To illustrate another type of energy we consider a pendulum (Fig. 4–7). If we pull the mass aside and release it, it swings back and forth. In its motion, it loses height in going from either end to the center. Where does the potential energy go? Gravitational energy disappears when it is down at the bottom; nevertheless, it will climb up again. The gravitational energy must have gone into another form. Evidently it is by virtue of its *motion* that it is able to climb up again, so we have the conversion of gravitational energy into some other form when it reaches the bottom.

We must get a formula for the energy of motion. Now, recalling our arguments about reversible machines, we can easily see that in the motion at the bottom must be a quantity of energy which permits it to rise a certain height, and which has nothing to do with the *machinery* by which it comes up or the *path* by which it comes up. So we have an equivalence formula something like the one we wrote for the child's blocks. We have another form to represent the energy. It is easy to say what it is. The kinetic energy at the bottom equals the weight times the height that it could go, corresponding to its velocity: K.E. $= WH$. What we need is the formula which tells us the height by some rule that has to do with the motion of objects. If we start something out with a certain

Conservation of Energy

Figure 4–7. Pendulum.

velocity, say straight up, it will reach a certain height; we do not know what it is yet, but it depends on the velocity—there is a formula for that. Then to find the formula for kinetic energy for an object moving with velocity V, we must calculate the height that it could reach, and multiply by the weight. We shall soon find that we can write it this way:

$$K.E. = WV^2/2g. \tag{4.6}$$

Of course, the fact that motion has energy has nothing to do with the fact that we are in a gravitational field. It makes no difference *where* the motion came from. This is a general formula for various velocities. Both (4.3) and (4.6) are approximate formulas, the first because it is incorrect when the heights are great, i.e., when the heights are so high that gravity is weakening; the second, because of the relativistic correction at high speeds. However, when we do finally get the exact formula for the energy, then the law of conservation of energy is correct.

Other forms of energy

We can continue in this way to illustrate the existence of energy in other forms. First, consider elastic energy. If we pull down on a spring, we must do some work, for when we have it down, we can lift weights with it. Therefore in its stretched condition it has a possibility of doing some work. If we were to evaluate the sums of weights times heights, it would not check out—we must add something else to account for the fact that the spring is under tension. Elastic energy is the formula for a spring when it is stretched. How much energy is it? If we let go, the elastic energy, as the spring passes through the equilibrium point, is converted to kinetic energy and it goes back and forth between compressing or stretching the spring and kinetic energy of motion. (There is also some gravitational energy going in and out, but we can do this experiment "sideways" if we like.) It keeps going until the losses—Aha! We have cheated all the way through by putting on little weights to move things or saying that the machines are reversible, or that they go on forever, but we can see that things do stop, eventually. Where is the energy when the spring has finished moving up and down? This brings in *another* form of energy: *heat energy.*

Inside a spring or a lever there are crystals which are made up of lots of atoms, and with great care and delicacy in the arrangement of the parts one can try to adjust things so that as something rolls on something else, none of the atoms do any jiggling at all. But one must be very careful. Ordinarily when things roll, there is bumping and jiggling because of the irregularities of the material, and the atoms start to wiggle inside. So we lose track of that energy; we find the atoms are wiggling inside in a random and confused manner after the motion slows down. There is still kinetic energy, all right, but it is not associated with visible motion. What a dream! How do we *know* there is still kinetic energy? It turns out that with thermometers you can find out that, in fact, the spring or the lever is *warmer*, and that there is really an increase of kinetic energy by a definite amount. We call this form of energy *heat energy*, but we know that it is not really a new form, it is just kinetic energy—internal motion. (One of the difficulties with all these experiments with matter that we do on a large scale is that we cannot really demonstrate the conservation of energy and we cannot really make our reversible machines, because every time we move a large clump of stuff, the atoms do not remain absolutely undisturbed, and so a

Conservation of Energy

certain amount of random motion goes into the atomic system. We cannot see it, but we can measure it with thermometers, etc.)

There are many other forms of energy, and of course we cannot describe them in any more detail just now. There is electrical energy, which has to do with pushing and pulling by electric charges. There is radiant energy, the energy of light, which we know is a form of electrical energy because light can be represented as wigglings in the electromagnetic field. There is chemical energy, the energy which is released in chemical reactions. Actually, elastic energy is, to a certain extent, like chemical energy, because chemical energy is the energy of the attraction of the atoms, one for the other, and so is elastic energy. Our modern understanding is the following: chemical energy has two parts, kinetic energy of the electrons inside the atoms, so part of it is kinetic, and electrical energy of interaction of the electrons and the protons—the rest of it, therefore, is electrical. Next we come to nuclear energy, the energy which is involved with the arrangement of particles inside the nucleus, and we have formulas for that, but we do not have the fundamental laws. We know that it is not electrical, not gravitational, and not purely chemical, but we do not know what it is. It seems to be an additional form of energy. Finally, associated with the relativity theory, there is a modification of the laws of kinetic energy, or whatever you wish to call it, so that kinetic energy is combined with another thing called *mass energy*. An object has energy from its sheer *existence*. If I have a positron and an electron, standing still doing nothing—never mind gravity, never mind anything—and they come together and disappear, radiant energy will be liberated, in a definite amount, and the amount can be calculated. All we need know is the mass of the object. It does not depend on what it is—we make two things disappear, and we get a certain amount of energy. The formula was first found by Einstein; it is $E = mc^2$.

It is obvious from our discussion that the law of conservation of energy is enormously useful in making analyses, as we have illustrated in a few examples without knowing all the formulas. If we had all the formulas for all kinds of energy, we could analyze how many processes should work without having to go into the details. Therefore conservation laws are very interesting. The question naturally arises as to what other conservation laws there are in physics. There are two other conservation laws which are analogous to the conservation of energy. One is called the conservation of linear momentum. The other is called the conservation of angular momentum. We will find out more about these later. In the last analysis, we do not understand the conservation laws deeply. We do not understand the conservation of energy. We do not understand energy as a certain number of little blobs. You may have heard that photons come out in blobs and that the energy of a photon is Planck's constant times the frequency. That is true, but since the frequency of light can be anything, there is no law that says that energy has to be a certain definite amount. Unlike Dennis' blocks, there can be any amount of energy, at least as presently understood. So we do not understand this energy as counting something at the moment, but just as a mathematical quantity, which is an abstract and rather peculiar circumstance. In quantum mechanics it turns out that the conservation of energy is very closely related to another important property of the world, *things do not depend on the absolute time.* We can set up an experiment at a given moment and try it out, and then do the same experiment at a later moment, and it will behave in exactly the same way. Whether this is strictly true or not, we do not know. If we assume that it *is* true, and add the principles of quantum mechanics, then we can deduce the principle of the conservation of energy. It is a rather subtle and interesting thing, and it is not easy to explain. The other conservation laws are also linked together. The conservation of momentum is associated in quantum mechanics with the proposition that it makes no difference *where* you do the experiment, the results will always be the same. As independence in space has to do with the conservation of momentum, independence of time has to do with the conservation of energy, and finally, if we *turn* our apparatus, this too makes no difference, and so the invariance of the world to angular orientation

Conservation of Energy

is related to the conservation of *angular momentum*. Besides these, there are three other conservation laws, that are exact so far as we can tell today, which *are* much simpler to understand because they are in the nature of counting blocks.

The first of the three is the *conservation of charge*, and that merely means that you count how many positive, minus how many negative electrical charges you have, and the number is never changed. You may get rid of a positive with a negative, but you do not create any net excess of positives over negatives. Two other laws are analogous to this one—one is called the *conservation of baryons*. There are a number of strange particles, a neutron and a proton are examples, which are called baryons. In any reaction whatever in nature, if we count how many baryons are coming into a process, the number of baryons* which come out will be exactly the same. There is another law, the *conservation of leptons*. We can say that the group of particles called leptons are: electron, mu meson, and neutrino. There is an antielectron which is a positron, that is, a −1 lepton. Counting the total number of leptons in a reaction reveals that the number in and out never changes, at least so far as we know at present.

These are the six conservation laws, three of them subtle, involving space and time, and three of them simple, in the sense of counting something.

With regard to the conservation of energy, we should note that *available* energy is another matter—there is a lot of jiggling around in the atoms of the water of the sea, because the sea has a certain temperature, but it is impossible to get them herded into a definite motion without taking energy from somewhere else. That is, although we know for a fact that energy is conserved, the energy available for human utility is not conserved so easily. The laws which govern how much energy is available are called the *laws of thermodynamics* and involve a concept called entropy for irreversible thermodynamic processes.

---

* Counting antibaryons as −1 baryon.

Finally, we remark on the question of where we can get our supplies of energy today. Our supplies of energy are from the sun, rain, coal, uranium, and hydrogen. The sun makes the rain, and the coal also, so that all these are from the sun. Although energy is conserved, nature does not seem to be interested in it; she liberates a lot of energy from the sun, but only one part in two billion falls on the earth. Nature has conservation of energy, but does not really care; she spends a lot of it in all directions. We have already obtained energy from uranium; we can also get energy from hydrogen, but at present only in an explosive and dangerous condition. If it can be controlled in thermonuclear reactions, it turns out that the energy that can be obtained from 10 quarts of water per second is equal to all of the electrical power generated in the United States. With 150 gallons of running water a minute, you have enough fuel to supply all the energy which is used in the United States today! Therefore it is up to the physicist to figure out how to liberate us from the need for having energy. It can be done.

Five

THE THEORY OF GRAVITATION

Planetary motions

In this chapter we shall discuss one of the most far-reaching generalizations of the human mind. While we are admiring the human mind, we should take some time off to stand in awe of a *nature* that could follow with such completeness and generality such an elegantly simple principle as the law of gravitation. What is this law of gravitation? It is that every object in the universe attracts every other object with a force which for any two bodies is proportional to the mass of each and varies inversely as the square of the distance between them. This statement can be expressed mathematically by the equation

$$F = G \frac{mm'}{r^2}.$$

If to this we add the fact that an object responds to a force by accelerating in the direction of the force by an amount that is inversely proportional to the mass of the object, we shall have said everything required, for a sufficiently talented mathematician could then deduce all the consequences of these two principles. However, since you are not assumed to be sufficiently talented yet, we shall discuss the consequences in more detail, and not just leave you with only these two bare principles. We shall briefly relate the story of the discovery of the law of gravitation and discuss some of its consequences, its effects on history, the mysteries that such a law entails, and some refinements of the law made by Einstein; we shall also discuss the relationships of the law to the other laws of physics. All this cannot be done in one chapter, but these subjects will be treated in due time in subsequent chapters.

The story begins with the ancients observing the motions of planets among the stars, and finally deducing that they went around the sun, a fact that was rediscovered later by Copernicus. Exactly *how* the planets went around the sun, with exactly *what motion*, took a little more work to discover. In the beginning of the fifteenth century there were great debates as to whether they really went around the sun or not. Tycho Brahe had an idea that was different from anything proposed by the ancients: his idea was that these debates about the nature of the motions of the planets would best be resolved if the actual positions of the planets in the sky were measured sufficiently accurately. If measurement showed exactly how the planets moved, then perhaps it would be possible to establish one or another viewpoint. This was a tremendous idea—that to find something out, it is better to perform some careful experiments than to carry on deep philosophical arguments. Pursuing this idea, Tycho Brahe studied the positions of the planets for many years in his observatory on the island of Hven, near Copenhagen. He made voluminous tables, which were then studied by the mathematician Kepler, after Tycho's death. Kepler discovered from the data some very beautiful and remarkable, but simple, laws regarding planetary motion.

Kepler's laws

First of all, Kepler found that each planet goes around the sun in a curve called an *ellipse*, with the sun at a focus of the ellipse. An ellipse is not just an oval, but is a very specific and precise curve that

The Theory of Gravitation

can be obtained by using two tacks, one at each focus, a loop of string, and a pencil; more mathematically, it is the locus of all points the sum of whose distances from two fixed points (the foci) is a constant. Or, if you will, it is a foreshortened circle (Fig. 5–1).

Kepler's second observation was that the planets do not go around the sun at a uniform speed, but move faster when they are nearer the sun and more slowly when they are farther from the sun, in precisely this way: Suppose a planet is observed at any two successive times, let us say a week apart, and that the radius vector* is drawn to the planet for each observed position. The orbital arc traversed by the planet during the week, and the two radius vectors, bound a certain plane area, the shaded area shown in Fig. 5–2. If two similar observations are made a week apart, at a part of the orbit farther from the sun (where the planet moves more slowly), the similarly bounded area is exactly the same as in the first case. So, in accordance with the second law, the orbital speed of each planet is such that the radius "sweeps out" equal areas in equal times.

Finally, a third law was discovered by Kepler much later; this law is of a different category from the other two, because it deals not with only a single planet, but relates one planet to another. This law says that when the orbital period and orbit size of any two planets are compared, the periods are proportional to the $\frac{3}{2}$ power of the

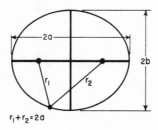

Figure 5–1. An ellipse.

---

* A radius vector is a line drawn from the sun to any point in a planet's orbit.

SIX EASY PIECES

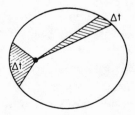

Figure 5–2. Kepler's law of areas.

orbit size. In this statement the period is the time interval it takes a planet to go completely around its orbit, and the size is measured by the length of the greatest diameter of the elliptical orbit, technically known as the major axis. More simply, if the planets went in circles, as they nearly do, the time required to go around the circle would be proportional to the ½ power of the diameter (or radius). Thus Kepler's three laws are:

   I. Each planet moves around the sun in an ellipse, with the sun at one focus.

  II. The radius vector from the sun to the planet sweeps out equal areas in equal intervals of time.

 III. The squares of the periods of any two planets are proportional to the cubes of the semimajor axes of their respective orbits: $T \sim a^{3/2}$.

Development of dynamics

While Kepler was discovering these laws, Galileo was studying the laws of motion. The problem was, what makes the planets go around? (In those days, one of the theories proposed was that the planets went around because behind them were invisible angels, beating their wings and driving the planets forward. You will see that this theory is now modified! It turns out that in order to keep the planets going around, the invisible angels must fly in a different

The Theory of Gravitation

direction and they have no wings. Otherwise, it is a somewhat similar theory!) Galileo discovered a very remarkable fact about motion, which was essential for understanding these laws. That is the principle of *inertia*—if something is moving, with nothing touching it and completely undisturbed, it will go on forever, coasting at a uniform speed in a straight line. (*Why* does it keep on coasting? We do not know, but that is the way it is.)

Newton modified this idea, saying that the only way to change the motion of a body is to use *force*. If the body speeds up, a force has been applied *in the direction of motion*. On the other hand, if its motion is changed to a new *direction*, a force has been applied *sideways*. Newton thus added the idea that a force is needed to change the speed *or the direction* of motion of a body. For example, if a stone is attached to a string and is whirling around in a circle, it takes a force to keep it in the circle. We have to *pull* on the string. In fact, the law is that the acceleration produced by the force is inversely proportional to the mass, or the force is proportional to the mass times the acceleration. The more massive a thing is, the stronger the force required to produce a given acceleration. (The mass can be measured by putting other stones on the end of the same string and making them go around the same circle at the same speed. In this way it is found that more or less force is required, the more massive object requiring more force.) The brilliant idea resulting from these considerations is that no *tangential* force is needed to keep a planet in its orbit (the angels do not have to fly tangentially) because the planet would coast in that direction anyway. If there were nothing at all to disturb it, the planet would go off in a *straight line*. But the actual motion deviates from the line on which the body would have gone if there were no force, the deviation being essentially *at right angles* to the motion, not in the direction of the motion. In other words, because of the principle of inertia, the force needed to control the motion of a planet around the sun is not a force *around* the sun but *toward* the sun. (If there is a force toward the sun, the sun might be the angel, of course!)

Newton's law of gravitation

From his better understanding of the theory of motion, Newton appreciated that *the sun* could be the seat or organization of forces that govern the motion of the planets. Newton proved to himself (and perhaps we shall be able to prove it soon) that the very fact that equal areas are swept out in equal times is a precise sign post of the proposition that all deviations are precisely *radial*—that the law of areas is a direct consequence of the idea that all of the forces are directed exactly *toward the sun*.

Next, by analyzing Kepler's third law it is possible to show that the farther away the planet, the weaker the forces. If two planets at different distances from the sun are compared, the analysis shows that the forces are inversely proportional to the squares of the respective distances. With the combination of the two laws, Newton concluded that there must be a force, inversely as the square of the distance, directed in a line between the two objects.

Being a man of considerable feeling for generalities, Newton supposed, of course, that this relationship applied more generally than just to the sun holding the planets. It was already known, for example, that the planet Jupiter had moons going around it as the moon of the earth goes around the earth, and Newton felt certain that each planet held its moons with a force. He already knew of the force holding *us* on the earth, so he proposed that this was a *universal* force—*that everything pulls everything else.*

The next problem was whether the pull of the earth on its people was the "same" as its pull on the moon, i.e., inversely as the square of the distance. If an object on the surface of the earth falls 16 feet in the first second after it is released from rest, how far does the moon fall in the same time? We might say that the moon does not fall at all. But if there were no force on the moon, it would go off in a straight line, whereas it goes in a circle instead, so it really *falls in* from where it would have been if there were no force at all. We can calculate from the radius of the moon's orbit (which is about 240,000 miles) and how long it takes to go around the earth (ap-

The Theory of Gravitation

proximately 29 days), how far the moon moves in its orbit in 1 second, and can then calculate how far it falls in one second.* This distance turns out to be roughly $\frac{1}{20}$ of an inch in a second. That fits very well with the inverse square law, because the earth's radius is 4000 miles, and if something which is 4000 miles from the center of the earth falls 16 feet in a second, something 240,000 miles, or 60 times as far away, should fall only $\frac{1}{3600}$ of 16 feet, which also is roughly $\frac{1}{20}$ of an inch. Wishing to put this theory of gravitation to a test by similar calculations, Newton made his calculations very carefully and found a discrepancy so large that he regarded the theory as contradicted by facts, and did not publish his results. Six years later a new measurement of the size of the earth showed that the astronomers had been using an incorrect distance to the moon. When Newton heard of this, he made the calculation again, with the corrected figures, and obtained beautiful agreement.

This idea that the moon "falls" is somewhat confusing, because, as you see, it does not come any *closer*. The idea is sufficiently interesting to merit further explanation: the moon falls in the sense that *it falls away from the straight line that it would pursue if there were no forces*. Let us take an example on the surface of the earth. An object released near the earth's surface will fall 16 feet in the first second. An object shot out *horizontally* will also fall 16 feet; even though it is moving horizontally, it still falls the same 16 feet in the same time. Figure 5–3 shows an apparatus which demonstrates this. On the horizontal track is a ball which is going to be driven forward a little distance away. At the same height is a ball which is going to fall vertically, and there is an electrical switch arranged so that at the moment the first ball leaves the track, the second ball is released. That they come to the same depth at the same time is witnessed by the fact that they collide in midair. An object like a bullet, shot horizontally, might go a long way in one second—perhaps 2000

---

* That is, how far the circle of the moon's orbit falls below the straight line tangent to it at the point where the moon was one second before.

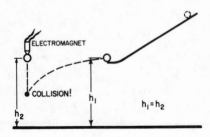

Figure 5–3. Apparatus for showing the independence of vertical and
horizontal motions.

feet—but it will still fall 16 feet if it is aimed horizontally. What happens if we shoot a bullet faster and faster? Do not forget that the earth's surface is curved. If we shoot it fast enough, then when it falls 16 feet it may be at just the same height above the ground as it was before. How can that be? It still falls, but the earth curves away, so it falls "around" the earth. The question is, how far does it have to go in one second so that the earth is 16 feet below the horizon? In Fig. 5–4 we see the earth with its 4000-mile radius, and the tangential, straightline path that the bullet would take if there were no force. Now, if we use one of those wonderful theorems in geometry, which says that our tangent is the mean proportional between the two parts of the diameter cut by an equal chord, we see that the horizontal distance travelled is the mean proportional between the 16 feet fallen and the 8000-mile diameter of the earth. The square root of $(16/5280) \times 8000$ comes out very close to 5 miles. Thus we see that if the bullet moves at 5 miles a second, it then will continue to fall toward the earth at the same rate of 16 feet each second, but will never get any closer because the earth keeps curving away from it. Thus it was that Mr. Gagarin maintained himself in space while going 25,000 miles around the earth at approximately 5 miles per second. (He took a little longer because he was a little higher.)

The Theory of Gravitation

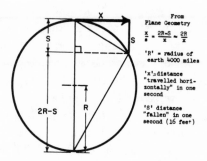

Figure 5–4. Acceleration toward the center of a circular path. From plane geometry, $x/s = (2R - S)/x \approx 2R/x$, where R is the radius of the earth, 4000 miles; x is the distance "travelled horizontally" in one second; and S is the distance "fallen" in one second (16 feet).

Any great discovery of a new law is useful only if we can take more out than we put in. Now, Newton *used* the second and third of Kepler's laws to deduce his law of gravitation. What did he *predict?* First, his analysis of the moon's motion was a prediction because it connected the falling of objects on the earth's surface with that of the moon. Second, the question is, *is the orbit an ellipse?* We shall see in a later chapter how it is possible to calculate the motion exactly, and indeed one can prove that it should be an ellipse,* so no extra fact is needed to explain Kepler's *first* law. Thus Newton made his first powerful prediction.

The law of gravitation explains many phenomena not previously understood. For example, the pull of the moon on the earth causes the tides, hitherto mysterious. The moon pulls the water up under it and makes the tides—people had thought of that before, but they were not as clever as Newton, and so they thought there ought to be only one tide during the day. The reasoning was that the moon pulls the water up under it, making a high tide and a low tide, and since

---

* The proof is not given in this course.

the earth spins underneath, that makes the tide at one station go up and down every 24 hours. Actually the tide goes up and down in 12 hours. Another school of thought claimed that the high tide should be on the other side of the earth because, so they argued, the moon pulls the earth away from the water! Both of these theories are wrong. It actually works like this: the pull of the moon for the earth and for the water is "balanced" at the center. But the water which is closer to the moon is pulled *more* than the average and the water which is farther away from it is pulled *less* than the average. Furthermore, the water can flow while the more rigid earth cannot. The true picture is a combination of these two things.

What do we mean by "balanced"? What balances? If the moon pulls the whole earth toward it, why doesn't the earth fall right "up" to the moon? Because the earth does the same trick as the moon, it goes in a circle around a point which is inside the earth but not at its center. The moon does not just go around the earth, the earth and the moon both go around a central position, each falling toward this common position, as shown in Fig. 5–5. This motion around the common center is what balances the fall of each. So the earth is not going in a straight line either; it travels in a circle. The water on the far side is "unbalanced" because the moon's attraction there is weaker than it is at the center of the earth, where it just balances the "centrifugal force." The result of this imbalance is that the water rises up, away from the center of the earth. On the near side, the attraction from the moon is stronger, and the imbalance is in the opposite direction in space, but again *away* from the center of the earth. The net result is that we get *two* tidal bulges.

Universal gravitation

What else can we understand when we understand gravity? Everyone knows the earth is round. Why is the earth round? That is easy; it is due to gravitation. The earth can be understood to be round merely because everything attracts everything else and so it has

The Theory of Gravitation

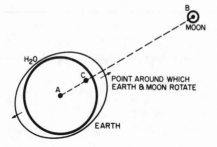

Figure 5–5. The earth-moon system, with tides.

attracted itself together as far as it can! If we go even further, the earth is not *exactly* a sphere because it is rotating, and this brings in centrifugal effects which tend to oppose gravity near the equator. It turns out that the earth should be elliptical, and we even get the right shape for the ellipse. We can thus deduce that the sun, the moon, and the earth should be (nearly) spheres, just from the law of gravitation.

What else can you do with the law of gravitation? If we look at the moons of Jupiter we can understand everything about the way they move around that planet. Incidentally, there was once a certain difficulty with the moons of Jupiter that is worth remarking on. These satellites were studied very carefully by Roemer, who noticed that the moons sometimes seemed to be ahead of schedule, and sometimes behind. (One can find their schedules by waiting a very long time and finding out how long it takes on the average for the moons to go around.) Now they were *ahead* when Jupiter was particularly *close* to the earth and they were *behind* when Jupiter was *farther* from the earth. This would have been a very difficult thing to explain according to the law of gravitation—it would have been, in fact, the death of this wonderful theory if there were no other explanation. If a law does not work even in *one place* where it ought to, it is just wrong. But the reason for this discrepancy was very simple and beautiful: it takes a little while to *see* the moons of Jupiter because of the time it takes light to travel from Jupiter to the earth.

When Jupiter is closer to the earth the time is a little less, and when it is farther from the earth, the time is more. This is why moons appear to be, on the average, a little ahead or a little behind, depending on whether they are closer to or farther from the earth. This phenomenon showed that light does not travel instantaneously, and furnished the first estimate of the speed of light. This was done in 1656.

If all of the planets push and pull on each other, the force which controls, let us say, Jupiter in going around the sun is not just the force from the sun; there is also a pull from, say, Saturn. This force is not really strong, since the sun is much more massive than Saturn, but there is *some* pull, so the orbit of Jupiter should not be a perfect ellipse, and it is not; it is slightly off, and "wobbles" around the correct elliptical orbit. Such a motion is a little more complicated. Attempts were made to analyze the motions of Jupiter, Saturn, and Uranus on the basis of the law of gravitation. The effects of each of these planets on each other were calculated to see whether or not the tiny deviations and irregularities in these motions could be completely understood from this one law. Lo and behold, for Jupiter and Saturn, all was well, but Uranus was "weird." It behaved in a very peculiar manner. It was not travelling in an exact ellipse, but that was understandable, because of the attractions of Jupiter and Saturn. But even if allowance were made for these attractions, Uranus *still* was not going right, so the laws of gravitation were in danger of being overturned, a possibility that could not be ruled out. Two men, Adams and Leverrier, in England and France, independently, arrived at another possibility: perhaps there is *another* planet, dark and invisible, which men had not seen. This planet, N, could pull on Uranus. They calculated where such a planet would have to be in order to cause the observed perturbations. They sent messages to the respective observatories, saying, "Gentlemen, point your telescope to such and such a place, and you will see a new planet." It often depends on with whom you are working as to whether they pay any attention to you or not. They did pay attention to Leverrier; they looked, and there planet N was! The other observatory then also looked very quickly in the next few days and saw it too.

The Theory of Gravitation

This discovery shows that Newton's laws are absolutely right in the solar system; but do they extend beyond the relatively small distances of the nearest planets? The first test lies in the question, do *stars* attract *each other* as well as planets? We have definite evidence that they do in the *double stars*. Figure 5–6 shows a double star—two stars very close together (there is also a third star in the picture so that we will know that the photograph was not turned). The stars are also shown as they appeared several years later. We see that, relative to the "fixed" star, the axis of the pair has rotated, i.e., the two stars are going around each other. Do they rotate according to Newton's laws? Careful measurements of the relative positions of one such double star system are shown in Fig. 5–7. There we see a beautiful ellipse, the measures starting in 1862 and going all the way around to 1904 (by now it must have gone around once more). Everything coincides with Newton's laws, except that the star Sirius A is *not at the focus*. Why should that be? Because the plane of the ellipse is not in the "plane of the sky." We are not looking at right angles to the orbit plane, and when an ellipse is viewed at a tilt, it remains an ellipse but the focus is no longer at the same place. Thus we can analyze double stars, moving about each other, according to the requirements of the gravitational law.

Figure 5–6. A double-star system.

SIX EASY PIECES

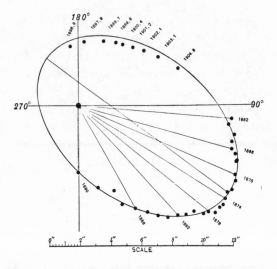

Figure 5–7. Orbit of Sirius *B* with respect to Sirius *A*.

That the law of gravitation is true at even bigger distances is indicated in Fig. 5–8. If one cannot see gravitation acting here, he has no soul. This figure shows one of the most beautiful things in the sky—a globular star cluster. All of the dots are stars. Although they look as if they are packed solid toward the center, that is due to the fallibility of our instruments. Actually, the distances between even the centermost stars are very great and they very rarely collide. There are more stars in the interior than farther out, and as we move outward there are fewer and fewer. It is obvious that there is an attraction among these stars. It is clear that gravitation exists at these enormous dimensions, perhaps 100,000 times the size of the solar system. Let us now go further, and look at an *entire galaxy*, shown in Fig. 5–9. The shape of this galaxy indicates an obvious tendency for its matter to agglomerate. Of course we cannot prove that the law here is precisely inverse square, only that there is still an attraction, at this enormous dimension, that holds the whole thing together. One may say, "Well, that is all very clever but why is it not just a ball?" Because it is *spinning* and has *angular momentum* which it cannot give up as it contracts; it must

The Theory of Gravitation

Figure 5–8. A globular star cluster.

contract mostly in a plane. (Incidentally, if you are looking for a good problem, the exact details of how the arms are formed and what determines the shapes of these galaxies has not been worked out.) It is, however, clear that the shape of the galaxy is due to gravitation even though the complexities of its structure have not yet allowed us to analyze it completely. In a galaxy we have a scale of perhaps 50,000 to 100,000 light years. The earth's distance from the sun is 8⅓ light *minutes*, so you can see how large these dimensions are.

Gravity appears to exist at even bigger dimensions, as indicated by Fig. 5–10, which shows many "little" things clustered together. This is a *cluster of galaxies*, just like a star cluster. Thus galaxies attract each other at such distances that they too are agglomerated into clusters. Perhaps gravitation exists even over distances of *tens of millions* of light years; so far as we now know, gravity seems to go out forever inversely as the square of the distance.

Not only can we understand the nebulae, but from the law of gravitation we can even get some ideas about the origin of the stars. If we have a big cloud of dust and gas, as indicated in Fig. 5–11, the gravitational attractions of the pieces of dust for one another might

Figure 5-9. A galaxy.

make them form little lumps. Barely visible in the figure are "little" black spots which may be the beginning of the accumulations of dust and gases which, due to their gravitation, begin to form stars. Whether we have ever seen a star form or not is still debatable. Figure 5-12 shows the one piece of evidence which suggests that we have. At the left is a picture of a region of gas with some stars in it taken in 1947, and at the right is another picture, taken only 7 years later, which shows two new bright spots. Has gas accumulated, has gravity acted hard enough and collected it into a ball big enough that the stellar nuclear reaction starts in the interior and turns it into a star? Perhaps, and perhaps not. It is unreasonable that in only seven years we should be so lucky as to see a star change itself into visible form; it is much less probable that we should see *two!*

Cavendish's experiment

Gravitation, therefore, extends over enormous distances. But if there is a force between *any* pair of objects, we ought to be able to measure the force between our own objects. Instead of having to

The Theory of Gravitation

Figure 5–10. A cluster of galaxies.

watch the stars go around each other, why can we not take a ball of lead and a marble and watch the marble go toward the ball of lead? The difficulty of this experiment when done in such a simple manner is the very weakness or delicacy of the force. It must be done with extreme care, which means covering the apparatus to keep the air out, making sure it is not electrically charged, and so on; then the force can be measured. It was first measured by Cavendish with an apparatus which is schematically indicated in Fig. 5–13. This first demonstrated the direct force between two large, fixed balls of lead and two smaller balls of lead on the ends of an arm supported by a very fine fiber, called a torsion fiber. By measuring how much the fiber gets twisted, one can measure the strength of the force, verify that it is inversely proportional to the square of the distance, and determine how strong it is. Thus, one may accurately determine the coefficient G in the formula

$$F = G\frac{mm'}{r^2}.$$

SIX EASY PIECES

Figure 5–11. An interstellar dust cloud.

All the masses and distances are known. You say, "We knew it already for the earth." Yes, but we did not know the *mass* of the earth. By knowing G from this experiment and by knowing how strongly the earth attracts, we can indirectly learn how great is the mass of the earth! This experiment has been called "weighing the earth." Cavendish claimed he was weighing the earth, but what he was measuring was the coefficient G of the gravity law. This is the only way in which the mass of the earth can be determined. G turns out to be

$$6.670 \times 10^{-11} \text{ newton} \cdot \text{m}^2/\text{kg}^2.$$

It is hard to exaggerate the importance of the effect on the history of science produced by this great success of the theory of gravitation. Compare the confusion, the lack of confidence, the incomplete knowledge that prevailed in the earlier ages, when there were endless debates and paradoxes, with the clarity and simplicity of this law—this fact that all the moons and planets and stars have

The Theory of Gravitation

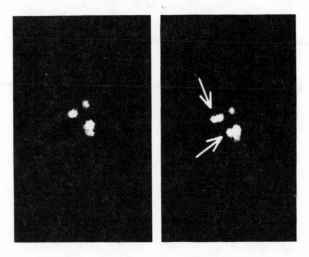

Figure 5–12. The formation of new stars?

such a *simple rule* to govern them, and further that man could *understand* it and deduce how the planets should move! This is the reason for the success of the sciences in following years, for it gave hope that the other phenomena of the world might also have such beautifully simple laws.

What is gravity?

But is this such a simple law? What about the machinery of it? All we have done is to describe *how* the earth moves around the sun, but we have not said *what makes it go*. Newton made no hypotheses about this; he was satisfied to find *what* it did without getting into the machinery of it. *No one has since given any machinery*. It is characteristic of the physical laws that they have this abstract character. The law of conservation of energy is a theorem concerning quantities that have to be calculated and added together, with no mention of the machinery, and likewise the great laws of mechanics are quantitative mathematical laws for which no machinery is available. Why can

SIX EASY PIECES

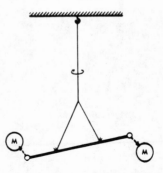

Figure 5–13. A simplified diagram of the apparatus used by Cavendish to verify the law of universal gravitation for small objects and to measure the gravitational constant G.

we use mathematics to describe nature without a mechanism behind it? No one knows. We have to keep going because we find out more that way.

Many mechanisms for gravitation have been suggested. It is interesting to consider one of these, which many people have thought of from time to time. At first, one is quite excited and happy when he "discovers" it, but he soon finds that it is not correct. It was first discovered about 1750. Suppose there were many particles moving in space at a very high speed in all directions and being only slightly absorbed in going through matter. When they *are* absorbed, they give an impulse to the earth. However, since there are as many going one way as another, the impulses all balance. But when the sun is nearby, the particles coming toward the earth through the sun are partially absorbed, so fewer of them are coming from the sun than are coming from the other side. Therefore, the earth feels a net impulse toward the sun and it does not take one long to see that it is inversely as the square of the distance—because of the variation of the solid angle that the sun subtends as we vary the distance. What is wrong with that machinery? It involves some new consequences which are *not true*. This particular idea has the

The Theory of Gravitation

following trouble: the earth, in moving around the sun, would impinge on more particles which are coming from its forward side than from its hind side (when you run in the rain, the rain in your face is stronger than that on the back of your head!). Therefore there would be more impulse given the earth from the front, and the earth would feel a *resistance to motion* and would be slowing up in its orbit. One can calculate how long it would take for the earth to stop as a result of this resistance, and it would not take long enough for the earth to still be in its orbit, so this mechanism does not work. No machinery has ever been invented that "explains" gravity without also predicting some other phenomenon that does *not* exist.

Next we shall discuss the possible relation of gravitation to other forces. There is no explanation of gravitation in terms of other forces at the present time. It is not an aspect of electricity or anything like that, so we have no explanation. However, gravitation and other forces are very similar, and it is interesting to note analogies. For example, the force of electricity between two charged objects looks just like the law of gravitation: the force of electricity is a constant, with a minus sign, times the product of the charges, and varies inversely as the square of the distance. It is in the opposite direction—likes repel. But is it still not very remarkable that the two laws involve the same function of distance? Perhaps gravitation and electricity are much more closely related than we think. Many attempts have been made to unify them; the so-called unified field theory is only a very elegant attempt to combine electricity and gravitation; but, in comparing gravitation and electricity, the most interesting thing is the *relative strengths* of the forces. Any theory that contains them both must also deduce how strong the gravity is.

If we take, in some natural units, the repulsion of two electrons (nature's universal charge) due to electricity, and the attraction of two electrons due to their masses, we can measure the ratio of electrical repulsion to the gravitational attraction. The ratio is independent of the distance and is a fundamental constant of nature.

The ratio is shown in Fig. 5–14. The gravitational attraction relative to the electrical repulsion between two electrons is 1 divided by 4.17 × 10^{42}! The question is, where does such a large number come from? It is not accidental, like the ratio of the volume of the earth to the volume of a flea. We have considered two natural aspects of the same thing, an electron. This fantastic number is a natural constant, so it involves something deep in nature. Where could such a tremendous number come from? Some say that we shall one day find the "universal equation," and in it, one of the roots will be this number. It is very difficult to find an equation for which such a fantastic number is a natural root. Other possibilities have been thought of; one is to relate it to the age of the universe. Clearly, we have to find *another* large number somewhere. But do we mean the age of the universe in *years?* No, because years are not "natural"; they were devised by men. As an example of something natural, let us consider the time it takes light to go across a proton, 10^{-24} second. If we compare this time with the *age of the universe*, 2×10^{10} years, the answer is 10^{-42}. It has about the same number of zeros going off it, so it has been proposed that the gravitational constant is related to the age of the universe. If that were the case, the gravitational constant would change with time, because as the

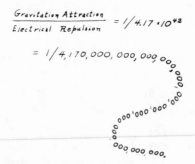

Figure 5–14. The relative strengths of electrical and gravitational interactions between two electrons.

The Theory of Gravitation

universe got older the ratio of the age of the universe to the time which it takes for light to go across a proton would be gradually increasing. Is it possible that the gravitational constant *is* changing with time? Of course the changes would be so small that it is quite difficult to be sure.

One test which we can think of is to determine what would have been the effect of the change during the past 10^9 years, which is approximately the age from the earliest life on the earth to now, and one-tenth of the age of the universe. In this time, the gravity constant would have increased by about 10 percent. It turns out that if we consider the structure of the sun—the balance between the weight of its material and the rate at which radiant energy is generated inside it—we can deduce that if the gravity were 10 percent stronger, the sun would be much more than 10 percent brighter— by the *sixth power* of the gravity constant! If we calculate what happens to the orbit of the earth when the gravity is changing, we find that the earth was then *closer in*. Altogether, the earth would be about 100 degrees centigrade hotter, and all of the water would not have been in the sea, but vapor in the air, so life would not have started in the sea. So we do *not* now believe that the gravity constant is changing with the age of the universe. But such arguments as the one we have just given are not very convincing, and the subject is not completely closed.

It is a fact that the force of gravitation is proportional to the *mass*, the quantity which is fundamentally a measure of *inertia*—of how hard it is to hold something which is going around in a circle. Therefore two objects, one heavy and one light, going around a larger object in the same circle at the same speed because of gravity, will stay together because to go in a circle *requires* a force which is stronger for a bigger mass. That is, the gravity is stronger for a given mass in *just the right proportion* so that the two objects will go around together. If one object were inside the other it would *stay* inside; it is a perfect balance. Therefore, Gagarin or Titov would find things "weightless" inside a space ship; if they happened to let go of a piece of chalk, for example, it would go around the earth in exactly the same way as the whole space ship, and so it would appear to remain suspended before them in space. It is very interesting that this force is *exactly* proportional to the mass with great precision, because if it were not exactly proportional there would be some effect by which inertia and weight would differ. The absence of such an effect has been checked with great accuracy by an experiment done first by Eötvös in 1909 and more recently by Dicke. For all substances tried, the masses and weights are exactly proportional within 1 part in 1,000,000,000, or less. This is a remarkable experiment.

Gravity and relativity

Another topic deserving discussion is Einstein's modification of Newton's law of gravitation. In spite of all the excitement it created, Newton's law of gravitation is not correct! It was modified by Einstein to take into account the theory of relativity. According to Newton, the gravitational effect is instantaneous, that is, if we were to move a mass, we would at once feel a new force because of the new position of that mass; by such means we could send signals at infinite speed. Einstein advanced arguments which suggest that we *cannot send signals faster than the speed of light*, so the law of gravitation must be wrong. By correcting it to take the delays into account, we have a new law, called Einstein's law of gravitation. One feature of this new law which is quite easy to understand is this: In the Einstein relativity theory, anything which has *energy* has mass—mass in the sense that it is attracted gravitationally. Even light, which has an energy, has a "mass." When a light beam, which has energy in it, comes past the sun there is an attraction on it by the sun. Thus the light does not go straight, but is deflected. During the eclipse of the sun, for example, the stars which are around the sun should appear displaced from where they would be if the sun were not there, and this has been observed.

Finally, let us compare gravitation with other theories. In recent

The Theory of Gravitation

years we have discovered that all mass is made of tiny particles and that there are several kinds of interactions, such as nuclear forces, etc. None of these nuclear or electrical forces has yet been found to explain gravitation. The quantum-mechanical aspects of nature have not yet been carried over to gravitation. When the scale is so small that we need the quantum effects, the gravitational effects are so weak that the need for a quantum theory of gravitation has not yet developed. On the other hand, for consistency in our physical theories it would be important to see whether Newton's law modified to Einstein's law can be further modified to be consistent with the uncertainty principle. This last modification has not yet been completed.

Six

QUANTUM BEHAVIOR

Atomic mechanics

In the last few chapters we have treated the essential ideas necessary for an understanding of most of the important phenomena of light—or electromagnetic radiation in general. (We have left a few special topics for next year. Specifically, the theory of the index of dense materials and total internal reflection.) What we have dealt with is called the "classical theory" of electric waves, which turns out to be a completely adequate description of nature for a large number of effects. We have not had to worry yet about the fact that light energy comes in lumps or "photons."

We would like to take up as our next subject the problem of the behavior of relatively large pieces of matter—their mechanical and thermal properties, for instance. In discussing these, we will find that the "classical" (or older) theory fails almost immediately, because matter is really made up of atomic-sized particles. Still, we will deal only with the classical part, because that is the only part that we can understand using the classical mechanics we have been learning. But we shall not be very successful. We shall find that in the case of matter, unlike the case of light, we shall be in difficulty relatively soon. We could, of course, continuously skirt away from the atomic effects, but we shall instead interpose here a short excursion in which we will describe the basic ideas of the quantum properties of matter, i.e., the quantum ideas of atomic physics, so that you will have some feeling for what it is we are leaving out. For we will have to leave out some important subjects that we cannot avoid coming close to.

So we will give now the *introduction* to the subject of quantum mechanics, but will not be able actually to get into the subject until much later.

"Quantum mechanics" is the description of the behavior of matter in all its details and, in particular, of the happenings on an atomic scale. Things on a very small scale behave like nothing that you have any direct experience about. They do not behave like waves, they do not behave like particles, they do not behave like clouds, or billiard balls, or weights on springs, or like anything that you have ever seen.

Newton thought that light was made up of particles, but then it was discovered, as we have seen here, that it behaves like a wave. Later, however (in the beginning of the twentieth century) it was found that light did indeed sometimes behave like a particle. Historically, the electron, for example, was thought to behave like a particle, and then it was found that in many respects it behaved like a wave. So it really behaves like neither. Now we have given up. We say: "It is like *neither*."

There is one lucky break, however—electrons behave just like light. The quantum behavior of atomic objects (electrons, protons, neutrons, photons, and so on) is the same for all; they are all "particle waves," or whatever you want to call them. So what we learn about the properties of electrons (which we shall use for our examples) will apply also to all "particles," including photons of light.

The gradual accumulation of information about atomic and small-scale behavior during the first quarter of this century, which gave some indications about how small things do behave, produced an increasing confusion which was finally resolved in 1926 and 1927 by Schrödinger, Heisenberg, and Born. They finally obtained a consistent description of the behavior of matter on a small scale. We take up the main features of that description in this chapter.

Quantum Behavior

Because atomic behavior is so unlike ordinary experience, it is very difficult to get used to and it appears peculiar and mysterious to everyone, both to the novice and to the experienced physicist. Even the experts do not understand it the way they would like to, and it is perfectly reasonable that they should not, because all of direct, human experience and of human intuition applies to large objects. We know how large objects will act, but things on a small scale just do not act that way. So we have to learn about them in a sort of abstract or imaginative fashion and not by connection with our direct experience.

In this chapter we shall tackle immediately the basic element of the mysterious behavior in its most strange form. We choose to examine a phenomenon which is impossible, *absolutely* impossible, to explain in any classical way, and which has in it the heart of quantum mechanics. In reality, it contains the *only* mystery. We cannot explain the mystery in the sense of "explaining" how it works. We will *tell* you how it works. In telling you how it works we will have told you about the basic peculiarities of all quantum mechanics.

An experiment with bullets

To try to understand the quantum behavior of electrons, we shall compare and contrast their behavior, in a particular experimental setup, with the more familiar behavior of particles like bullets, and with the behavior of waves like water waves. We consider first the behavior of bullets in the experimental setup shown diagrammatically in Fig. 6–1. We have a machine gun that shoots a stream of bullets. It is not a very good gun, in that it sprays the bullets (randomly) over a fairly large angular spread, as indicated in the figure. In front of the gun we have a wall (made of armor plate) that has in it two holes just about big enough to let a bullet through. Beyond the wall is a backstop (say a thick wall of wood) which will "absorb" the bullets when they hit it. In front of the wall we have an object which we shall call a "detector" of bullets. It might be a box

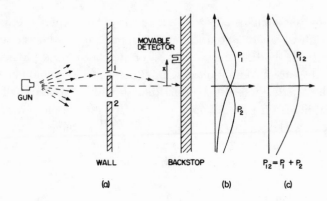

Figure 6–1. Interference experiment with bullets.

containing sand. Any bullet that enters the detector will be stopped and accumulated. When we wish, we can empty the box and count the number of bullets that have been caught. The detector can be moved back and forth (in what we will call the x-direction). With this apparatus, we can find out experimentally the answer to the question: "What is the probability that a bullet which passes through the holes in the wall will arrive at the backstop at the distance x from the center?" First, you should realize that we should talk about probability, because we cannot say definitely where any particular bullet will go. A bullet which happens to hit one of the holes may bounce off the edges of the hole, and may end up anywhere at all. By "probability" we mean the chance that the bullet will arrive at the detector, which we can measure by counting the number which arrive at the detector in a certain time and then taking the ratio of this number to the *total* number that hit the backstop during that time. Or, if we assume that the gun always shoots at the same rate during the measurements, the probability we want is just proportional to the number that reach the detector in some standard time interval.

For our present purposes we would like to imagine a somewhat idealized experiment in which the bullets are not real bullets, but

Quantum Behavior

are *indestructible* bullets—they cannot break in half. In our experiment we find that bullets always arrive in lumps, and when we find something in the detector, it is always one whole bullet. If the rate at which the machine gun fires is made very low, we find that at any given moment either nothing arrives, or one and only one—exactly one—bullet arrives at the backstop. Also, the size of the lump certainly does not depend on the rate of firing of the gun. We shall say: "Bullets *always* arrive in identical lumps." What we measure with our detector is the probability of arrival of a lump. And we measure the probability as a function of x. The result of such measurements with this apparatus (we have not yet done the experiment, so we are really imagining the result) is plotted in the graph drawn in part (c) of Fig. 6–1. In the graph we plot the probability to the right and x vertically, so that the x-scale fits the diagram of the apparatus. We call the probability P_{12} because the bullets may have come either through hole 1 or through hole 2. You will not be surprised that P_{12} is large near the middle of the graph but gets small if x is very large. You may wonder, however, why P_{12} has its maximum value at $x = 0$. We can understand this fact if we do our experiment again after covering up hole 2, and once more while covering up hole 1. When hole 2 is covered, bullets can pass only through hole 1, and we get the curve marked P_1 in part (b) of the figure. As you would expect, the maximum of P_1 occurs at the value of x which is on a straight line with the gun and hole 1. When hole 1 is closed, we get the symmetric curve P_2 drawn in the figure. P_2 is the probability distribution for bullets that pass through hole 2. Comparing parts (b) and (c) of Fig. 6–1, we find the important result that

$$P_{12} = P_1 + P_2. \tag{6.1}$$

The probabilities just add together. The effect with both holes open is the sum of the effects with each hole open alone. We shall call this result an observation of "*no interference*," for a reason that you will see later. So much for bullets. They come in lumps, and their probability of arrival shows no interference.

An experiment with waves

Now we wish to consider an experiment with water waves. The apparatus is shown diagrammatically in Fig. 6–2. We have a shallow trough of water. A small object labeled the "wave source" is jiggled up and down by a motor and makes circular waves. To the right of the source we have again a wall with two holes, and beyond that is a second wall, which, to keep things simple, is an "absorber," so that there is no reflection of the waves that arrive there. This can be done by building a gradual sand "beach." In front of the beach we place a detector which can be moved back and forth in the x-direction, as before. The detector is now a device which measures the "intensity" of the wave motion. You can imagine a gadget which measures the height of the wave motion, but whose scale is calibrated in proportion to the *square* of the actual height, so that the reading is proportional to the intensity of the wave. Our detector reads, then, in proportion to the *energy* being carried by the wave—or rather, the rate at which energy is carried to the detector.

With our wave apparatus, the first thing to notice is that the intensity can have *any* size. If the source just moves a very small

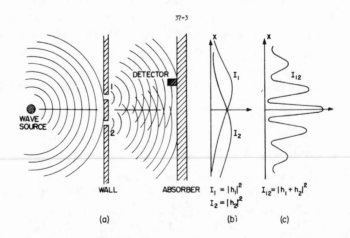

37-3

Figure 6–2. Interference experiment with water waves.

Quantum Behavior

amount, then there is just a little bit of wave motion at the detector. When there is more motion at the source, there is more intensity at the detector. The intensity of the wave can have any value at all. We would *not* say that there was any "lumpiness" in the wave intensity.

Now let us measure the wave intensity for various values of x (keeping the wave source operating always in the same way). We get the interesting-looking curve marked I_{12} in part (c) of the figure.

We have already worked out how such patterns can come about when we studied the interference of electric waves. In this case we would observe that the original wave is diffracted at the holes, and new circular waves spread out from each hole. If we cover one hole at a time and measure the intensity distribution at the absorber we find the rather simple intensity curves shown in part (b) of the figure. I_1 is the intensity of the wave from hole 1 (which we find by measuring when hole 2 is blocked off) and I_2 is the intensity of the wave from hole 2 (seen when hole 1 is blocked).

The intensity I_{12} observed when both holes are open is certainly *not* the sum of I_1 and I_2. We say that there is "interference" of the two waves. At some places (where the curve I_{12} has its maxima) the waves are "in phase" and the wave peaks add together to give a large amplitude and, therefore, a large intensity. We say that the two waves are "interfering constructively" at such places. There will be such constructive interference wherever the distance from the detector to one hole is a whole number of wavelengths larger (or shorter) than the distance from the detector to the other hole.

At those places where the two waves arrive at the detector with a phase difference of π (where they are "out of phase") the resulting wave motion at the detector will be the difference of the two amplitudes. The waves "interfere destructively," and we get a low value for the wave intensity. We expect such low values wherever the distance between hole 1 and the detector is different from the distance between hole 2 and the detector by an odd number of half-wavelengths. The low values of I_{12} in Fig. 6–2 correspond to the places where the two waves interfere destructively.

You will remember that the quantitative relationship between I_1,

I_2, and I_{12} can be expressed in the following way: The instantaneous height of the water wave at the detector for the wave from hole 1 can be written as (the real part of) $\hat{h}_1 e^{i\omega t}$, where the "amplitude" $\hat{h}_1$ is, in general, a complex number. The intensity is proportional to the mean squared height or, when we use the complex numbers, to $|\hat{h}_1|^2$. Similarly, for hole 2 the height is $\hat{h}_2 e^{i\omega t}$ and the intensity is proportional to $|\hat{h}_2|^2$. When both holes are open, the wave heights add to give the height $(\hat{h}_1 + \hat{h}_2) e^{i\omega t}$ and the intensity $|\hat{h}_1 + \hat{h}_2|^2$. Omitting the constant of proportionality for our present purposes, the proper relations for *interfering waves* are

$$I_1 = |\hat{h}_1|^2, \qquad I_2 = |\hat{h}_2|^2, \qquad I_{12} = |\hat{h}_1 + \hat{h}_2|^2. \qquad (6.2)$$

You will notice that the result is quite different from that obtained with bullets (Eq. 6.1). If we expand $|\hat{h}_1 + \hat{h}_2|^2$ we see that

$$|\hat{h}_1 + \hat{h}_2|^2 = |\hat{h}_1|^2 + |\hat{h}_2|^2 + 2|\hat{h}_1||\hat{h}_2| \cos \delta, \qquad (6.3)$$

where δ is the phase difference between $\hat{h}_1$ and $\hat{h}_2$. In terms of the intensities, we could write

$$I_{12} = I_1 + I_2 + 2\sqrt{I_1 I_2} \cos \delta. \qquad (6.4)$$

The last term in (6.4) is the "interference term." So much for water waves. The intensity can have any value, and it shows inteference.

An experiment with electrons

Now we imagine a similar experiment with electrons. It is shown diagrammatically in Fig. 6–3. We make an electron gun which consists of a tungsten wire heated by an electric current and surrounded by a metal box with a hole in it. If the wire is at a negative voltage with respect to the box, electrons emitted by the wire will be accelerated toward the walls and some will pass through the hole. All the electrons which come out of the gun will have (nearly) the same energy.

Quantum Behavior

In front of the gun is again a wall (just a thin metal plate) with two holes in it. Beyond the wall is another plate which will serve as a "backstop." In front of the backstop we place a movable detector. The detector might be a geiger counter or, perhaps better, an electron multiplier, which is connected to a loudspeaker.

We should say right away that you should not try to set up this experiment (as you could have done with the two we have already described). This experiment has never been done in just this way. The trouble is that the apparatus would have to be made on an impossibly small scale to show the effects we are interested in. We are doing a "thought experiment," which we have chosen because it is easy to think about. We know the results that *would* be obtained because there *are* many experiments that have been done, in which the scale and the proportions have been chosen to show the effects we shall describe.

The first thing we notice with our electron experiment is that we hear sharp "clicks" from the detector (that is, from the loudspeaker). And all "clicks" are the same. There are *no* "half-clicks."

We would also notice that the "clicks" come very erratically. Something like: click . . . . . click-click . . . click . . . . click . . . .

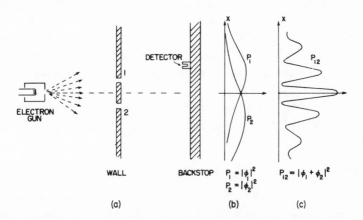

Figure 6–3. Interference experiment with electrons.

click-click . . . . . . click . . . , etc., just as you have, no doubt, heard a geiger counter operating. If we count the clicks which arrive in a sufficiently long time—say for many minutes—and then count again for another equal period, we find that the two numbers are very nearly the same. So we can speak of the *average rate* at which the clicks are heard (so-and-so-many clicks per minute on the average).

As we move the detector around, the *rate* at which the clicks appear is faster or slower, but the size (loudness) of each click is always the same. If we lower the temperature of the wire in the gun the rate of clicking slows down, but still each click sounds the same. We would notice also that if we put two separate detectors at the backstop, one *or* the other would click, but never both at once. (Except that once in a while, if there were two clicks very close together in time, our ear might not sense the separation.) We conclude, therefore, that whatever arrives at the backstop arrives in "lumps." All the "lumps" are the same size: only whole "lumps" arrive, and they arrive one at a time at the backstop. We shall say: "Electrons always arrive in identical lumps."

Just as for our experiment with bullets, we can now proceed to find experimentally the answer to the question: "What is the relative probability that an electron 'lump' will arrive at the backstop at various distances x from the center?" As before, we obtain the relative probability by observing the rate of clicks, holding the operation of the gun constant. The probability that lumps will arrive at a particular x is proportional to the average rate of clicks at that x.

The result of our experiment is the interesting curve marked P_{12} in part (c) of Fig. 6–3. Yes! That is the way electrons go.

The interference of electron waves

Now let us try to analyze the curve of Fig. 6–3 to see whether we can understand the behavior of the electrons. The first thing we would say is that since they come in lumps, each lump, which we may as

Quantum Behavior

well call an electron, has come either through hole 1 or through hole 2. Let us write this in the form of a "Proposition":

Proposition A: Each electron *either* goes through hole 1 *or* it goes through hole 2.

Assuming Proposition A, all electrons that arrive at the backstop can be divided into two classes: (1) those that come through hole 1, and (2) those that come through hole 2. So our observed curve must be the sum of the effects of the electrons which come through hole 1 and the electrons which come through hole 2. Let us check this idea by experiment. First, we will make a measurement for those electrons that come through hole 1. We block off hole 2 and make our counts of the clicks from the detector. From the clicking rate, we get P_1. The result of the measurement is shown by the curve marked P_1 in part (b) of Fig. 6–3. The result seems quite reasonable. In a similar way, we measure P_2, the probability distribution for the electrons that come through hole 2. The result of this measurement is also drawn in the figure.

The result P_{12} obtained with *both* holes open is clearly not the sum of P_1 and P_2, the probabilities for each hole alone. In analogy with our water-wave experiment, we say: "There is interference."

$$\textit{For electrons:} \qquad P_{12} \neq P_1 + P_2. \qquad (6.5)$$

How can such an interference come about? Perhaps we should say: "Well, that means, presumably, that it is *not true* that the lumps go either through hole 1 or hole 2, because if they did, the probabilities should add. Perhaps they go in a more complicated way. They split in half and . . ." But no! They cannot, they always arrive in lumps . . . "Well, perhaps some of them go through 1, and then they go around through 2, and then around a few more times, or by some other complicated path . . . then by closing hole 2, we changed the chance that an electron that *started out* through hole 1 would finally get to the backstop . . ." But notice! There are some points at which very few electrons arrive when *both* holes are open, but which receive many electrons if we close one hole, so *closing* one hole *increased* the number from the other. Notice, however, that at the center of the pattern, P_{12} is more than twice as large as $P_1 + P_2$. It is as though closing one hole *decreased* the number of electrons which come through the other hole. It seems hard to explain *both* effects by proposing that the electrons travel in complicated paths.

It is all quite mysterious. And the more you look at it the more mysterious it seems. Many ideas have been concocted to try to explain the curve for P_{12} in terms of individual electrons going around in complicated ways through the holes. None of them has succeeded. None of them can get the right curve for P_{12} in terms of P_1 and P_2.

Yet, surprisingly enough, the *mathematics* for relating P_1 and P_2 to P_{12} is extremely simple. For P_{12} is just like the curve I_{12} of Fig. 6–2, and *that* was simple. What is going on at the backstop can be described by two complex numbers that we can call $\hat{\phi}_1$ and $\hat{\phi}_2$ (they are functions of x, of course). The absolute square of $\hat{\phi}_1$ gives the effect with only hole 1 open. That is, $P_1 = |\hat{\phi}_1|^2$. The effect with only hole 2 open is given by $\hat{\phi}_2$ in the same way. That is, $P_2 = |\hat{\phi}_1|^2$. And the combined effect of the two holes is just $P_{12} = |\hat{\phi}_1 + \hat{\phi}_2|^2$. The *mathematics* is the same as what we had for the water waves! (It is hard to see how one could get such a simple result from a complicated game of electrons going back and forth through the plate on some strange trajectory.)

We conclude the following: The electrons arrive in lumps, like particles, and the probability of arrival of these lumps is distributed like the distribution of intensity of a wave. It is in this sense that an electron behaves "sometimes like a particle and sometimes like a wave."

Incidentally, when we were dealing with classical waves we defined the intensity as the mean over time of the square of the wave amplitude, and we used complex numbers as a mathematical trick to simplify the analysis. But in quantum mechanics it turns out that the amplitudes *must* be represented by complex numbers. The real

Quantum Behavior

parts alone will not do. That is a technical point, for the moment, because the formulas look just the same.

Since the probability of arrival through both holes is given so simply, although it is not equal to $(P_1 + P_2)$, that is really all there is to say. But there are a large number of subtleties involved in the fact that nature does work this way. We would like to illustrate some of these subtleties for you now. First, since the number that arrives at a particular point is *not* equal to the number that arrives through 1 plus the number that arrives through 2, as we would have concluded from Proposition A, undoubtedly we should conclude that *Proposition A is false*. It is *not* true that the electrons go *either* through hole 1 or hole 2. But that conclusion can be tested by another experiment.

Watching the electrons

We shall now try the following experiment. To our electron apparatus we add a very strong light source, placed behind the wall and between the two holes, as shown in Fig. 6–4. We know that electric charges scatter light. So when an electron passes, however it does pass, on its way to the detector, it will scatter some light to our eye, and we can *see* where the electron goes. If, for instance, an electron were to take the path via hole 2 that is sketched in Fig. 6–4, we should see a flash of light coming from the vicinity of the place marked A in the figure. If an electron passes through hole 1 we would expect to see a flash from the vicinity of the upper hole. If it should happen that we get light from both places at the same time, because the electron divides in half . . . Let us just do the experiment!

Here is what we see: *every* time that we hear a "click" from our electron detector (at the backstop), we *also see* a flash of light *either* near hole 1 *or* near hole 2, but *never* both at once! And we observe the same result no matter where we put the detector. From this observation we conclude that when we look at the electrons we find that the electrons go either through one hole or the other. Experimentally, Proposition A is necessarily true.

SIX EASY PIECES

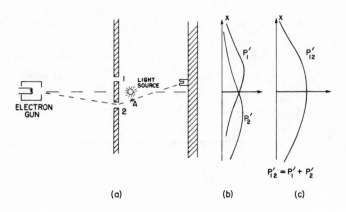

Figure 6–4. A different electron experiment.

What, then, is wrong with our argument *against* Proposition A? Why *isn't* P_{12} just equal to $P_1 + P_2$? Back to experiment! Let us keep track of the electrons and find out what they are doing. For each position (x-location) of the detector we will count the electrons that arrive and *also* keep track of which hole they went through, by watching for the flashes. We can keep track of things this way: whenever we hear a "click" we will put a count in Column 1 if we see the flash near hole 1, and if we see the flash near hole 2, we will record a count in Column 2. Every electron which arrives is recorded in one of two classes: those which come through 1 and those which come through 2. From the number recorded in Column 1 we get the probability P'_1 that an electron will arrive at the detector via hole 1; and from the number recorded in Column 2 we get P'_2, the probability that an electron will arrive at the detector via hole 2. If we now repeat such a measurement for many values of x, we get the curves for P'_1 and P'_2 shown in part (b) of Fig. 6–4.

Well, that is not too surprising! We get for P'_1 something quite similar to what we got before for P_1 by blocking off hole 2; and P'_2 is similar to what we got by blocking hole 1. So there is *not* any complicated business like going through both holes. When we watch them, the electrons come through just as we would expect

Quantum Behavior

them to come through. Whether the holes are closed or open, those which we see come through hole 1 are distributed in the same way whether hole 2 is open or closed.

But wait! What do we have *now* for the *total* probability, the probability that an electron will arrive at the detector by any route? We already have that information. We just pretend that we never looked at the light flashes, and we lump together the detector clicks which we have separated into the two columns. We *must* just *add* the numbers. For the probability that an electron will arrive at the backstop by passing through *either* hole, we do find $P'_{12} = P_1 + P_2$. That is, although we succeeded in watching which hole our electrons come through, we no longer get the old interference curve P_{12}, but a new one, P'_{12}, showing no interference! If we turn out the light P_{12} is restored.

We must conclude that *when we look at the electrons* the distribution of them on the screen is different than when we do not look. Perhaps it is turning on our light source that disturbs things? It must be that the electrons are very delicate, and the light, when it scatters off the electrons, gives them a jolt that changes their motion. We know that the electric field of the light acting on a charge will exert a force on it. So perhaps we *should* expect the motion to be changed. Anyway, the light exerts a big influence on the electrons. By trying to "watch" the electrons we have changed their motions. That is, the jolt given to the electron when the photon is scattered by it is such as to change the electron's motion enough so that if it *might* have gone to where P_{12} was at a maximum it will instead land where P_{12} was a minimum; that is why we no longer see the wavy interference effects.

You may be thinking: "Don't use such a bright source! Turn the brightness down! The light waves will then be weaker and will not disturb the electrons so much. Surely, by making the light dimmer and dimmer, eventually the wave will be weak enough that it will have a negligible effect." O.K. Let's try it. The first thing we observe is that the flashes of light scattered from the electrons as they pass by does *not* get weaker. *It is always the same-sized flash.* The only thing that happens as the light is made dimmer is that sometimes we hear a "click" from the detector but see *no flash at all*. The electron has gone by without being "seen." What we are observing is that light *also* acts like electrons; we *knew* that it was "wavy," but now we find that it is also "lumpy." It always arrives—or is scattered—in lumps that we call "photons." As we turn down the *intensity* of the light source we do not change the *size* of the photons, only the *rate* at which they are emitted. *That* explains why, when our source is dim, some electrons get by without being seen. There did not happen to be a photon around at the time the electron went through.

This is all a little discouraging. If it is true that whenever we "see" the electron we see the same-sized flash, then those electrons we see are *always* the disturbed ones. Let us try the experiment with a dim light anyway. Now whenever we hear a click in the detector we will keep a count in three columns: in Column (1) those electrons seen by hole 1, in Column (2) those electrons seen by hole 2, and in Column (3) those electrons not seen at all. When we work up our data (computing the probabilities) we find these results: Those "seen by hole 1" have a distribution like P'_1; those "seen by hole 2" have a distribution like P'_2 (so that those "seen by either hole 1 or 2" have a distribution like P'_{12}); and those "not seen at all" have a "wavy" distribution just like P_{12} of Fig. 6–3! *If the electrons are not seen, we have interference!*

That is understandable. When we do not see the electron, no photon disturbs it, and when we do see it, a photon has disturbed it. There is always the same amount of disturbance because the light photons all produce the same-sized effects and the effect of the photons being scattered is enough to smear out any interference effect.

Is there not *some* way we can see the electrons without disturbing them? We learned in an earlier chapter that the momentum carried by a "photon" is inversely proportional to its wavelength ($p = h/\lambda$). Certainly the jolt given to the electron when the photon is scattered toward our eye depends on the momentum that photon carries.

Quantum Behavior

Aha! If we want to disturb the electrons only slightly we should not have lowered the *intensity* of the light, we should have lowered its *frequency* (the same as increasing its wavelength). Let us use light of a redder color. We could even use infrared light, or radiowaves (like radar), and "see" where the electron went with the help of some equipment that can "see" light of these longer wavelengths. If we use "gentler" light perhaps we can avoid disturbing the electrons so much.

Let us try the experiment with longer waves. We shall keep repeating our experiment, each time with light of a longer wavelength. At first, nothing seems to change. The results are the same. Then a terrible thing happens. You remember that when we discussed the microscope we pointed out that, due to the *wave nature* of the light, there is a limitation on how close two spots can be and still be seen as two separate spots. This distance is of the order of the wavelength of light. So now, when we make the wavelength longer than the distance between our holes, we see a *big* fuzzy flash when the light is scattered by the electrons. We can no longer tell which hole the electron went through! We just know it went somewhere! And it is just with light of this color that we find that the jolts given to the electron are small enough so that P'_{12} begins to look like P_{12}—that we begin to get some interference effect. And it is only for wavelengths much longer than the separation of the two holes (when we have no chance at all of telling where the electron went) that the disturbance due to the light gets sufficiently small that we again get the curve P_{12} shown in Fig. 6–3.

In our experiment we find that it is impossible to arrange the light in such a way that one can tell which hole the electron went through, and at the same time not disturb the pattern. It was suggested by Heisenberg that the then new laws of nature could only be consistent if there were some basic limitation on our experimental capabilities not previously recognized. He proposed, as a general principle, his *uncertainty principle*, which we can state in terms of our experiment as follows: "It is impossible to design an apparatus to determine which hole the electron passes through, that will not at the same time disturb the electrons enough to destroy the interference pattern." If an apparatus is capable of determining which hole the electron goes through, it *cannot* be so delicate that it does not disturb the pattern in an essential way. No one has ever found (or even thought of) a way around the uncertainty principle. So we must assume that it describes a basic characteristic of nature.

The complete theory of quantum mechanics which we now use to describe atoms and, in fact, all matter depends on the correctness of the uncertainty principle. Since quantum mechanics is such a successful theory, our belief in the uncertainty principle is reinforced. But if a way to "beat" the uncertainty principle were ever discovered, quantum mechanics would give inconsistent results and would have to be discarded as a valid theory of nature.

"Well," you say, "what about Proposition A? It is true, or is it *not* true, that the electron either goes through hole 1 or it goes through hole 2?" The only answer that can be given is that we have found from experiment that there is a certain special way that we have to think in order that we do not get into inconsistencies. What we must say (to avoid making wrong predictions) is the following: If one looks at the holes or, more accurately, if one has a piece of apparatus which is capable of determining whether the electrons go through hole 1 or hole 2, then one *can* say that it goes either through hole 1 or hole 2. *But*, when one does *not* try to tell which way the electron goes, when there is nothing in the experiment to disturb the electrons, then one may *not* say that an electron goes either through hole 1 or hole 2. If one does say that, and starts to make any deductions from the statement, he will make errors in the analysis. This is the logical tightrope on which we must walk if we wish to describe nature successfully.

◊   ◊   ◊

If the motion of all matter—as well as electrons—must be described in terms of waves, what about the bullets in our first experiment? Why didn't we see an interference pattern there? It turns out that for the bullets the wavelengths were so tiny that the

Quantum Behavior

interference patterns became very fine. So fine, in fact, that with any detector of finite size one could not distinguish the separate maxima and minima. What we saw was only a kind of average, which is the classical curve. In Fig. 6–5 we have tried to indicate schematically what happens with large-scale objects. Part (a) of the figure shows the probability distribution one might predict for bullets, using quantum mechanics. The rapid wiggles are supposed to represent the interference pattern one gets for waves of very short wave-length. Any physical detector, however, straddles several wiggles of the probability curve, so that the measurements show the smooth curve drawn in part (b) of the figure.

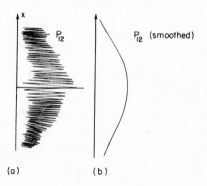

Figure 6–5. Interference pattern with bullets: (a) actual (schematic),
(b) observed.

First principles of quantum mechanics

We will now write a summary of the main conclusions of our experiments. We will, however, put the results in a form which makes them true for a general class of such experiments. We can write our summary more simply if we first define an "ideal experiment" as one in which there are no uncertain external influences, i.e., no jiggling or other things going on that we cannot take into account. We would be quite precise if we said: "An ideal experiment is one in which all of the initial and final conditions of the experiment are completely specified." What we will call "an event" is, in general, just a specific set of initial and final conditions. (For example: "an electron leaves the gun, arrives at the detector, and nothing else happens.") Now for our summary.

SUMMARY

(1) The probability of an event in an ideal experiment is given by the square of the absolute value of a complex number ø which is called the probability amplitude.

$$P = \text{probability},$$
$$\phi = \text{probability amplitude}, \qquad (6.6)$$
$$P = |\phi|^2.$$

(2) When an event can occur in several alternative ways, the probability amplitude for the event is the sum of the probability amplitudes for each way considered separately. There is interference.

$$\phi = \phi_1 + \phi_2,$$
$$P = |\phi_1 + \phi_2|^2. \qquad (6.7)$$

(3) If an experiment is performed which is capable of determining whether one or another alternative is actually taken, the probability of the event is the sum of the probabilities for each alternative. The interference is lost.

$$P = P_1 + P_2. \qquad (6.8)$$

One might still like to ask: "How does it work? What is the machinery behind the law?" No one has found any machinery behind the law. No one can "explain" any more than we have just "explained." No one will give you any deeper representation of the situation. We have no ideas about a more basic mechanism from which these results can be deduced.

We would like to emphasize a very important difference between classical and quantum mechanics. We have been talking about the probability that an electron will arrive in a given circumstance. We have implied that in our experimental arrangement (or even in the best possible one) it would be impossible to predict exactly what would happen. We can only predict the odds! This would mean, if it were true, that physics has given up on the problem of trying to predict exactly what will happen in a definite circumstance. Yes! Physics *has* given up. *We do not know how to predict what would happen in a given circumstance*, and we believe now that it is impossible, that the only thing that can be predicted is the probability of different events. It must be recognized that this is a retrenchment in our earlier ideal of understanding nature. It may be a backward step, but no one has seen a way to avoid it.

We make now a few remarks on a suggestion that has sometimes been made to try to avoid the description we have given: "Perhaps the electron has some kind of internal works—some inner variables—that we do not yet know about. Perhaps that is why we cannot predict what will happen. If we could look more closely at the electron we could be able to tell where it would end up." So far as we know, that is impossible. We would still be in difficulty. Suppose we were to assume that inside the electron there is some kind of machinery that determines where it is going to end up. That machine must *also* determine which hole it is going to go through on its way. But we must not forget that what is inside the electron should not be dependent on what *we* do, and in particular upon whether we open or close one of the holes. So if an electron, before it starts, has already made up its mind (a) which hole it is going to use, and (b) where it is going to land, we should find P_1 for those electrons that have chosen hole 1, P_2 for those that have chosen hole 2, *and necessarily* the sum $P_1 + P_2$ for those that arrive through the two holes. There seems to be no way around this. But we have verified experimentally that that is not the case. And no one has figured a way out of this puzzle. So at the present time we must limit ourselves to computing probabilities. We say "at the present time," but we suspect very strongly that it is something that will be with us forever—that it is impossible to beat that puzzle—that this is the way nature really *is*.

The uncertainty principle

This is the way Heisenberg stated the uncertainty principle originally: If you make the measurement on any object, and you can determine the x-component of its momentum with an uncertainty Δp, you cannot, at the same time, know its x-position more accurately than $\Delta x = h/\Delta p$. The uncertainties in the position and momentum at any instant must have their product greater than Planck's constant. This is a special case of the uncertainty principle that was stated above more generally. The more general statement was that one cannot design equipment in any way to determine which of two alternatives is taken, without, at the same time, destroying the pattern of interference.

Let us show for one particular case that the kind of relation given by Heisenberg must be true in order to keep from getting into trouble. We imagine a modification of the experiment of Fig. 6–3, in which the wall with the holes consists of a plate mounted on rollers so that it can move freely up and down (in the x-direction), as shown in Fig. 6–6. By watching the motion of the plate carefully we can try to tell which hole an electron goes through. Imagine what happens when the detector is placed at $x = 0$. We would expect that an electron which passes through hole 1 must be deflected downward by the plate to reach the detector. Since the vertical component of the electron momentum is changed, the plate must recoil with an equal momentum in the opposite direction. The plate will get an upward kick. If the electron goes through the lower hole, the plate should feel a downward kick. It is clear that for every position of the detector, the momentum received by the plate will have a

Quantum Behavior

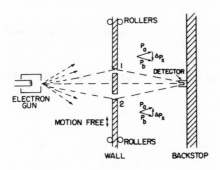

Figure 6–6. An experiment in which the recoil of the wall is measured.

different value for a traversal via hole 1 than for a traversal via hole 2. So! Without disturbing the electrons *at all*, but just by watching the *plate*, we can tell which path the electron used.

Now in order to do this it is necessary to know what the momentum of the screen is, before the electron goes through. So when we measure the momentum after the electron goes by, we can figure out how much the plate's momentum has changed. But remember, according to the uncertainty principle we cannot at the same time know the position of the plate with an arbitrary accuracy. But if we do not know exactly *where* the plate is we cannot say precisely where the two holes are. They will be in a different place for every electron that goes through. This means that the center of our interference pattern will have a different location for each electron. The wiggles of the interference pattern will be smeared out. We shall show quantitatively in the next chapter that if we determine the momentum of the plate sufficiently accurately to determine from the recoil measurement which hole was used, then the uncertainty in the *x*-position of the plate will, according to the uncertainty principle, be enough to shift the pattern observed at the detector up and down in the *x*-direction about the distance from a maximum to its nearest minimum. Such a random shift is just enough to smear out the pattern so that no interference is observed.

The uncertainty principle "protects" quantum mechanics. Heisenberg recognized that if it were possible to measure the momentum and the position simultaneously with a greater accuracy, the quantum mechanics would collapse. So he proposed that it must be impossible. Then people sat down and tried to figure out ways of doing it, and nobody could figure out a way to measure the position and the momentum of anything—a screen, an electron, a billiard ball, anything—with any greater accuracy. Quantum mechanics maintains its perilous but accurate existence.

Index

Acceleration, force, mass and, 89, 93
Acceleration toward the center of a
    circular path, 97f
Acetylcholine, 51
Activation energy, 52
Actomyosin, 51
Adenine, 57
Air
    turbulent flow, 62
    water evaporation, 11–13
α-irone, 19f
Amalgamation, classes of natural
    phenomena, 26–27
Amino acids, 55, 57, 59
Angstrom (A), 5
Angular momentum, 84, 85, 102
Antineutrons, 38
Antiparticles, 38, 42
Antiprotons, 38
Astronomy, physics and, 59–61, 65
Atmospheric pressure, 6
Atomic bomb, 38
Atomic fact, 4
Atomic hypothesis, 4–10
Atomic mechanics, quantum behavior,
    115–17
Atoms. *See also* Quantum mechanics
    chemical reactions, 15–21
    as components of all matter, 4–10
    evidence for, 19–21
    introduction, 1–4
    labeling, 54–55

physical processes and, 10–15
properties, 27
size, 5, 34
Attraction, electrical force and, 28
Available energy, 85

Baryons, 40, 42, 44, 85
Basic physics. *See also* Physics
Beta-decay particle interactions, 43–
    44
Biology, physics and, 49–59
Brahe, Tycho, 90
Brownian motion, defined, 20
Bullets, quantum behavior
    experiment, 117–19
Burning, defined, 16

Carbon, 15–16
Carbon dioxide (CO_2), 16, 17
Carbon monoxide, 16
Caughey, Prof. T., *xxvi*
Cavendish's experiment, 104–7, 108f
Cells, biological functions, 50–53
Challenger (space shuttle), *xi*
Character of Physical Law, The (book),
    viii
Charge
    conservation of, 85
    electric fields and, 28–31
    of the electron, 38
    quantum electrodynamics, 37
Chemical energy, 71, 83

Chemical formula, defined, 18
Chemical reactions
  atomic hypothesis, 15–21
  as natural phenomenon, 26
Chemistry, physics and, 48–49
Circulating fluids, 66
Cluster of galaxies, 103
Compression, of a gas, 8
Conservation of energy
  gravitational potential energy, 72–80
  kinetic energy, 80–81
  law of, 50, 69–72
  other forms of energy, 81–86
Copernicus, 90
Cosmic rays, 32t, 33, 38
Coupling, photons, 43–44
Crystalline array, 9
Curved space-time, 33
Cytosine, 57

Density
  gas pressure proportional to, 8
  waves of excess, 28
Desoxyribose nucleic acid (DNA),
    56–59
Dissolving air and solids in water, 12–
    15
DNA (desoxyribose nucleic acid), 56–
    59
Double stars, 101
Dynamic processes, defined, 14

Earth, mass of, 106
Earth-moon system, with tides, 97–98,
    99f
Earth sciences, 61
Earthquakes, 62–63
Einstein, Albert, *xi*, 33, 83
Elastic energy, 71, 81–82
Electric field
  defined, 30
  magnetism and, 31
  quantum electrodynamics, 37
Electrical energy, 71, 83

Electrical particle interactions, 43–44
Electrical potential energy, 76–77
Electricity
  compared to gravitation, 29, 109–10
  forces of, 28–30
  as natural phenomenon, 26
Electromagnetic field, defined, 26, 31
Electromagnetic waves, 31–33, 32t
Electron microscopes, 19
Electrons. *See also* Quantum behavior
  compared, muons, 43
  determine chemical properties, 30
  mass/charge, 38
  positrons as opposite charge, 37–38
  properties of, 29–30
Ellipse, 90, 91f, 97
E=mc², 83
Energy. *See also* Conservation of
    energy
  formulas for differing forms of, 70–
    72
  gravity and, 43
  of motion, 80
  of the stars/sun, 60–61
  supplies, 85–86
Entropy, 85
Enzymes, 53–57
Erosion, 62
Euclid, 27
Evaporation, 11–13
Evolution, 65
Expansion, of a gas, 8, 10
Experimentation. *See also* Quantum
    behavior
  scientific method and, 24, 35–36
  techniques, physics and, 54–55

Ferments, 53
Feynman, Richard, *vii-viii, xxvi*
Feynman Lectures of Physics, *xi-xv*
Force
  inertia and, 27–28, 93
  proportional to area, 7
  strength of, distance and, 103–7

Frequencies, electromagnetic waves, 31–33, 32t, 36–37
Fundamental physics, 24–25, 36

Galaxy, shape of, 102–3
Galileo, 92
Gamma rays, 32–33, 37–38
Gases, atomic hypothesis, 6–10
GDP and GTP, 54
Gell-Mann, Murray, *vii*
Geology, physics and, 61–63, 65, 66
Globular star cluster, 102
Goodstein, David L., *xv*
Gravitation
   Cavendish's experiment, 104–7, 108f
   compared to electricity, 29, 109–10
   curved space-time, 33
   defined, 27
   development of dynamics, 92–93
   Kepler's laws, 90–92
   mechanisms of, 107–12
   as natural phenomenon, 26
   Newton's law, 94–98
   particle interactions, 43–44
   planetary motions, 89–90
   relativity and, 112–13
   universal, 98–104
Gravitational energy, 71, 80
Gravitational potential energy, 72–80
Gravitons, 40, 43
Guanadine-di-phosphate (GDP), 54
Guanadine-tri-phosphate (GTP), 54
Guanine, 57

Heat
   from burning, 16
   molecular motion and, 5–6
   moving particles, 28
   as natural phenomenon, 26
   statistical mechanics and, 48–49
Heat energy, 71, 82
Helium, 10, 60, 61
Hemoglobin, 56

Horizontal motions, 95, 96f
Hydrogen, 61

Ice, atomic hypothesis, 9–10
Imino acid, 55
Inclined plane, smooth, 77
Index of dense materials, theory of, 115
Inertia, 27, 93, 111–12
Infrared frequencies, 32
Inorganic chemistry, 48, 49
Interaction force, 27
Interference experiments, quantum behavior, 117–33, 134
Inverse square law, 94–95, 102
Ions, defined, 13
Irreversible thermodynamic processes, entropy for, 85
Isotopes, 55, 61

Jupiter, moons of, 99–100

K-mesons, 42
Kepler's laws, 90–92
Kinetic energy, 34, 71, 80–82
Krebs cycle, 52–54

Lambda baryons, 41, 42
Laws, of natural phenomena, 26, 69, 89–90, 92–93
Leighton, Prof. R.B., *xxvi*
Leptons, 40, 42, 43, 85
Light
   astronomy, physics and, 59–61
   electrons, quantum behavior experiment with, 127–33
   emission, 37–38
   energy of, 83, 115
   generating, burning and, 16
   as natural phenomenon, 26
   quantum electrodynamics, 37
Light microscopes, 19
Light waves, 31, 116
Linear momentum, 84

Liquids, atomic hypothesis, 9
Long-range interaction force, 27

Machines, weight-lifting, 72–76
Magnetism
    charges in relative motion, 30–31
    as natural phenomenon, 26
Manhattan Project (Los Alamos), *vii*
Mass
    earth, 106
    electrons, 38
    proportional force of gravitation,
        111–12
Mass energy, 71, 83
Mathematics, physics and, 47
Matter. *See also* Quantum behavior
    atomic hypothesis, 4–10
    particles in, 27–28
    quantum electrodynamics, 37
Mechanics, as natural phenomenon,
    26
Melting, defined, 10
Membrane, cell, 50
Memory, 64
Mendeléev periodic table, 39–40, 48
Meson, natural phenomena of, 26
Mesons, 40, 42, 44
Meteorology, 61–62
Mev unit, 40
Microscopes, 19
Microsomes, 58, 59
Molecules
    chemical reactions, 15–21
    concept of, 14, 15
    oxygen/nitrogen, 11, 12
Momentum
    conservation of, 84, 85
    uncertainty of, 34
Moon, law of gravitation and, 94–95,
    97–98
Motion
    energy of, 80–81
    laws of, 92–93
Mountains, 62–63

μ-meson, 38–39, 43
Muons, 38–39, 43
Myosin, 51

Natural philosophy, 47
Nature, phenomena of, 26
Nebulae, law of gravitation and,
    103–4
Neher, Prof. H.V., *xxv, xxvi*
Nerve cells, 50–51
Nervous system, 64
Neugebauer, Prof. G., *xv, xxvi*
Neutrino, 43
Neutrons, 29–30, 38, 40
Newton's laws, 33, 94–98, 101, 112–13
Nitrogen molecules, 11, 12, 17
Nobel Prize, Physics, *vii, xi, xiv*
Nuclear energy, 71, 83
Nuclear force particle interaction, 43–44
Nuclear physics
    as natural phenomenon, 26
    nuclei, particles and, 38–45
Nuclear reactions, stars, 61
Nuclei
    particles and, 29, 38–45
    size, 34

Observation, scientific method and, 24
Odor, molecules and, 17–18
Optics, as natural phenomenon, 26
Organic chemistry, 18, 49
Oscillatory waves, 31–33, 32t
Oxygen molecules, 11, 12, 15–17

Particles
    behavior of, 33, 36, 116
    defined, 27
    elementary, 41t
    interaction types, 43–44
    masses at rest, 43
    nuclei and, 38–45
    "strangeness" number, 40
    types, 28, 29–30
Pasteur, Louis, 66

Pendulum, 80
Periodic table of elements, 39–40, 48
Perpetual motion, 72
Philosophy, physics and, 35–36, 47
Photons
    coupling, 43–44
    electrons, quantum behavior
        experiment with light, 130–31
    energy of, 84
    as opposite charge to electrons, 37–38
    properties of, 40
Photosynthesis, 52
Physical chemistry, 49
Physics
    before 1920, 27–33
    astronomy and, 59–61
    biology and, 49–59
    chemistry and, 48–49
    geology and, 61–63
    introduction, 23–27
    nuclear, 26, 38–45
    nuclei and particles, 38–45
    psychology and, 63–64
    relationship to other sciences, 47, 64–67
π-meson, 39
Pions, 39, 42
Planck's constant, 84, 136
Planetary motions, gravitation and, 89–90
Plants
    biology of, 51–52
    transpiration, 17
Plesset, Prof. M., *xxvi*
Position, uncertainty of, 34
Positrons, 37–38
Potential energy, 76
Precipitation, 14
Pressure
    atmospheric, 6
    confining a gas, 6–8
Principle of virtual work, 79
Probability, 118–19, 134

Prolene, 55–56
Proteins, 53, 55–59
Protons, 29–30, 38, 40
Psychology, physics and, 63–64

Q.E.D.: The Strange Theory of Light and Matter (book), *viii*
Quantum behavior
    atomic mechanics, 115–17
    bullets, experiment with, 117–19
    electron waves, interference of, 124–27
    electrons, experiment with, 122–24, 127–33
    waves, experiment with, 120–22
Quantum chemistry, 49
Quantum electrodynamics, 37–38, 43–44
Quantum mechanics
    of chemistry, 26, 48
    defined, 116
    first principles of, 133–36
    gravitation and, 113
    properties of, 33–38
    uncertainty principle, 136–38

Radar waves, 32
Radiant energy, 71, 83
Radio waves, 31
Radius vector, 91
Reason, scientific method and, 24
Relativity
    gravity and, 112–13
    theory of, 43
Reproduction, 57–59
Repulsion, electrical force and, 28
Resonances, 40, 42
Respiratory cycle, 52
Reversible machines, 73–76
RNA, 58–59
"Rules of the game," 24–25

S number, particles, 40
Salt, dissolving in water, 13–15

Sands, Prof. Matthew, *xx, xxiv, xxvi*
Schwinger, Julian, *vii, xiv*
Scientific method, 24
Screw jack, 78
Sensation, physiology of, 64
Short-range interaction force, 28
Sigma baryons, 41, 42
Sodium chloride, 13
Solids
    atomic hypothesis, 9
    dissolving in water, 13–15
Solution, precipitation and, 14
Sound, waves of excess density and, 28
Space, conservation of momentum, 84, 85
Space, time and, 27, 33, 44
Stabler, Prof. H.P., *xxvi*
Stars, 59–61, 65–66, 101–4
Statistical mechanics, 48–49, 60
Steam, atomic hypothesis, 6–8
Steam engines, 71
Stevinus, 78
"Strangeness" number, particles, 40, 42
Strong, Prof. F., *xxvi*
Strong particle interactions, 43–44
Substances, properties of, 26
Sun, planetary motion and, 90–92
Sutton, Prof. R.M., *xxvi*
Symmetry, atoms in ice, 9

Technetium, 60
Temperature. *See also* Heat
    of a gas, 8–10
    geology and, 62–63
Thermodynamics, 49, 85
    ͏nuclear reactions, 86

    ͏d, 97–98, 99f
    ͏ergy and, 84,

    3

Tomanaga, Sin-Itero, *vii, xiv*
Torsion fiber, 105
Total internal reflection, 115
Transpiration, 17
Turbulent flow, 62
Turbulent fluids, 66

Ultraviolet frequencies, 32
Unbalanced electrical force, 29
Uncertainty principle, 34, 131–32, 136–38
Universal gravitation, 98–104
Universe, age of, 110–11

Velocity, kinetic energy and, 81
Vertical motions, 95, 96f
Virtual work, principle of, 79
Vogt, Prof. R., *xxvi*
Vulcanism, 62–63

Water
    atomic hypothesis, 4–10
    dissolving solids in, 13–15
    evaporation, 11–13
Water vapor, 11–13, 17
Waves
    behavior of, 32–33, 36
    electromagnetic, 31–33
    excess density, sound and, 28
    quantum behavior experiment with, 120–22
Weak decay particle interactions, 43–44
Weather, 61–62, 66
Weight-lifting machines, 72–76
Wilts, Prof. C.H., *xxvi*
Work, virtual, 79

X-rays, 26, 32–33

Zero-charge particles, 40
Zero mass, 43
Zero-mass particles, 40

About Richard Feynman

Born in 1918 in Brooklyn, Richard P. Feynman received his Ph.D. from Princeton in 1942. Despite his youth, he played an important part in the Manhattan Project at Los Alamos during World War II. Subsequently, he taught at Cornell and at the California Institute of Technology. In 1965 he received the Nobel Prize in Physics, along with Sin-Itero Tomanaga and Julian Schwinger, for his work in quantum electrodynamics.

Dr. Feynman won his Nobel Prize for successfully resolving problems with the theory of quantum electrodynamics. He also created a mathematical theory that accounts for the phenomenon of superfluidity in liquid helium. Thereafter, with Murray Gell-Mann, he did fundamental work in the area of weak interactions such as beta decay. In later years Feynman played a key role in the development of quark theory by putting forward his parton model of high energy proton collision processes.

Beyond these achievements, Dr. Feynman introduced basic new computational techniques and notations into physics, above all, the ubiquitous Feynman diagrams that, perhaps more than any other formalism in recent scientific history, have changed the way in which basic physical processes are conceptualized and calculated.

Feynman was a remarkably effective educator. Of all his numerous awards, he was especially proud of the Oersted Medal for Teaching which he won in 1972. *The Feynman Lectures on Physics*, originally published in 1963, were described by a reviewer in *Scientific American* as "tough, but nourishing and full of flavor. After years it is *the* guide for teachers and for the best of begi students." In order to increase the understanding of physics among the lay public, Dr. Feynman wrote *The Character of Physical Law* and *Q.E.D.: The Strange Theory of Light and Matter*. He also authored a number of advanced publications that have become classic references and textbooks for researchers and students.

Richard Feynman was a constructive public man. His work on the Challenger commission is well known, especially his famous demonstration of the susceptibility of the O-rings to cold, an elegant experiment which required nothing more than a glass of ice water. Less well known were Dr. Feynman's efforts on the California State Curriculum Committee in the 1960's where he protested the mediocrity of textbooks.

A recital of Richard Feynman's myriad scientific and educational accomplishments cannot adequately capture the essence of the man. As any reader of even his most technical publications knows, Feynman's lively and multi-sided personality shines through all his work. Besides being a physicist, he was at various times a repairer of radios, a picker of locks, an artist, a dancer, a bongo player, and even a decipherer of Mayan Hieroglyphics. Perpetually curious about his world, he was an exemplary empiricist.

Richard Feynman died on February 15, 1988, in Los Angeles.

SIX EASY PIECES
AND
SIX NOT-SO-EASY PIECES

SIX
NOT-SO-EASY
PIECES

SIX
NOT-SO-EASY
PIECES

*Einstein's
Relativity, Symmetry,
and Space-Time*

RICHARD P.
FEYNMAN

*Originally prepared for publication by
Robert B. Leighton and Matthew Sands*

New Introduction by Roger Penrose

Contents

Publisher's Note   *vii*
Introduction by Roger Penrose   *ix*
Special Preface   *xvii*
Feynman's Preface   *xxiii*

ONE: Vectors   1

1-1 Symmetry in physics   1
1-2 Translations   2
1-3 Rotations   5
1-4 Vectors   9
1-5 Vector algebra   12
1-6 Newton's laws in vector notation   15
1-7 Scalar product of vectors   18

TWO: Symmetry in Physical Laws   23

2-1 Symmetry operations   23
2-2 Symmetry in space and time   24
2-3 Symmetry and conservation laws   29
2-4 Mirror reflections   30
2-5 Polar and axial vectors   35
2-6 Which hand is right?   38
2-7 Parity is not conserved!   40
2-8 Antimatter   43
2-9 Broken symmetries   46

THREE: The Special Theory of Relativity   49

3-1 The principle of relativity   49
3-2 The Lorentz transformation   53

3-3 The Michelson-Morley experiment   54

3-4 Transformation of time   59

3-5 The Lorentz contraction   63

3-6 Simultaneity   63

3-7 Four-vectors   65

3-8 Relativistic dynamics   66

3-9 Equivalence of mass and energy   68

FOUR: Relativistic Energy and Momentum   73

4-1 Relativity and the philosophers   73

4-2 The twin paradox   77

4-3 Transformation of velocities   79

4-4 Relativistic mass   83

4-5 Relativistic energy   88

FIVE: Space-Time   93

5-1 The geometry of space-time   93

5-2 Space-time intervals   97

5-3 Past, present, and future   99

5-4 More about four-vectors   102

5-5 Four-vector algebra   106

SIX: Curved Space   111

6-1 Curved spaces with two dimensions   111

6-2 Curvature in three-dimensional space   123

6-3 Our space is curved   125

6-4 Geometry in space-time   128

6-5 Gravity and the principle of equivalence   129

6-6 The speed of clocks in a gravitational field   131

6-7 The curvature of space-time   137

6-8 Motion in curved space-time   137

6-9 Einstein's theory of gravitation   141

Index   145

About Richard Feynman   151

PUBLISHER'S NOTE

The unqualified success and popularity of SIX EASY PIECES (Perseus Publishing, 1995) sparked a clamor, from the general public, students, and professional scientists alike, for *more* Feynman in book and audio. So we went back to the original *Lectures on Physics* and to the Archives at Caltech to see if there were more " easy" pieces. There were not. But there were many *not-so-easy* lectures that, although they contain some mathematics, are not too difficult for beginning science students; and for the student *and* the lay person, these six lectures are very bit as thrilling, as absorbing, and as much fun as the first six.

Another difference between these not-so-easy pieces and the first six, is that the topics of the first six spanned several fields of physics, from mechanics to thermodynamics to atomic physics. These new six pieces you hold in your hand, however, are focused around a subject which has evoked many of the most revolutionary discoveries and amazing theories of modern physics, from black holes to worm holes, from atomic energy to time warps; we are talking, of course, about Relativity. But even the great Einstein himself, the father of Relativity, could not *explain* the wonders, workings, and fundamental concepts of his own theory as well as could that guy from Noo Yawk, Richard P. Feynman, as reading the chapters or listening to CDs will prove to you.

Publisher's Note

Perseus Publishing wishes to thank Roger Penrose for his penetrating Introduction to this collection; Brian Hatfield and David Pines for their invaluable advice in the selection of the six lectures; and the California Institute of Technology's Physics Department and Institute Archives, in particular Judith Goodstein, for helping make this book/CD project happen.

Introduction

To understand why Richard Feynman was such a great teacher, it is important to appreciate his remarkable stature as a scientist. He was indeed one of the outstanding figures of twentieth-century theoretical physics. His contributions to that subject are central to the whole development of the particular way in which quantum theory is used in current cutting-edge research and thus to our present-day pictures of the world. The Feynman path integrals, Feynman diagrams, and Feynman rules are among the very basic tools of the modern theoretical physicist—tools that are necessary for the application of the rules of quantum theory to physical fields (e.g., the quantum theory of electrons, protons, and photons), and which form an essential part of the procedures whereby one makes these rules consistent with the requirements of Einstein's Special Relativity theory. Although none of these ideas is easy to appreciate, Feynman's particular approach always had a deep clarity about it, sweeping away unnecessary complications in what had gone before. There was a close link between his special ability to make progress in research and his particular qualities as a teacher. He had a unique talent that enabled him to cut through the complications that often obscure the essentials of a physical issue and to see clearly into the deep underlying physical principles.

Yet, in the popular conception of Feynman, he is known more for his antics and buffoonery, for his practical jokes, his irreverence towards authority, his bongo-drum performing, his relationships with women, both deep and shallow, his attendance at strip clubs, his attempts, late in life, to reach the obscure country of Tuva in central Asia, and many other schemes. Undoubtedly, he must have been extraordinarily clever, as his lightning quickness at calculation, his exploits involving safe-cracking, outwitting security services, deciphering ancient Mayan texts—not to mention his eventual Nobel Prize—clearly demonstrate. Yet none of this quite conveys the status that he unquestionably has amongst physicists and other scientists, as one of the deepest and most original thinkers of this century.

The distinguished physicist and writer Freeman Dyson, an early collaborator of Feynman's at a time when he was developing his most important ideas, wrote in a letter to his parents in England in the spring of 1948, when Dyson was a graduate student at Cornell University, "Feynman is the young American professor, half genius and half buffoon, who keeps all physicists and their children amused with his effervescent vitality. He has, however, as I have recently learned, a great deal more to him than that. . . ." Much later, in 1988, he would write: "A truer description would have said that Feynman was all genius and all buffoon. The deep thinking and the joyful clowning were not separate parts of a split personality. . . . He was thinking and clowning simultaneously."* Indeed, in his lectures, his wit was spontaneous, and often outrageous. Through it he held his audiences' attention, but never in a way that would distract from the purpose of the lecture, which was the conveying of genuine and deep physical understanding. Through laughter, his audiences could relax and be at ease, rather than feel daunted by what might otherwise be somewhat intimidating mathematical expressions and physical concepts that are tantalizingly difficult to grasp. Yet, although he enjoyed being center stage and was undoubtedly a showman, this was not the purpose of his exposi-

---

* The Dyson quotations are to be found in his book *From Eros to Gaia* (Pantheon Books, New York, 1992) pages 325 and 314, respectively.

tions. That purpose was to convey some basic understanding of underlying physical ideas and of the essential mathematical tools that are needed in order to express these ideas properly.

Whereas laughter played a key part of his success in holding an audience's attention, more important to the conveying of understanding was the immediacy of his approach. Indeed, he had an extraordinarily direct no-nonsense style. He scorned airy-fairy philosophizing where it had little physical content. Even his attitude to mathematics was somewhat similar. He had little use for pedantic mathematical niceties, but he had a distinctive mastery of the mathematics that he needed, and could present it in a powerfully transparent way. He was beholden to no one, and would never take on trust what others might maintain to be true without himself coming to an independent judgment. Accordingly, his approach was often strikingly original whether in his research or teaching. And when Feynman's way differed significantly from what had gone before, it would be a reasonably sure bet that Feynman's approach would be the more fruitful one to follow.

Feynman's preferred method of communication was verbal. He did not easily, nor often, commit himself to the printed word. In his scientific papers, the special "Feynman" qualities would certainly come through, though in a somewhat muted form. It was in his lectures that his talents were given full reign. His exceedingly popular "Feynman Lectures" were basically edited transcripts (by Robert B. Leighton and Matthew Sands) of lectures that Feynman gave, and the compelling nature of the text is evident to anyone who reads it. The *SIX NOT-SO-EASY PIECES* that are presented here are taken from those accounts. Yet, even here, the printed words alone leave something significantly missing. To sense the full excitement that Feynman's lectures exude, I believe that it is important to hear his actual voice. The directness of Feynman's approach, the irreverence, and the humor then become things that we can immediately share in. Fortunately, there are recordings of all the lectures presented in this book, which give us this opportunity—and I strongly recommend, if the opportunity is there, that at least some

Introduction

of these audio versions are listened to first. Once we have heard Feynman's forceful, enthralling, and witty commentary, in the tones of this streetwise New Yorker, we do not forget how he sounds, and it gives us an image to latch on to when we read his words. But whether we actually read the chapters or not, we can share something of the evident thrill that he himself feels as he explores—and continually re-explores—the extraordinary laws that govern the workings of our universe.

The present series of six lectures was carefully chosen to be of a level a little above the six that formed the earlier set of Feynman lectures entitled *Six Easy Pieces* (published by Addison Wesley Longman in 1995). Moreover, they go well together and constitute a superb and compelling account of one of the most important general areas of modern theoretical physics.

This area is *relativity,* which first burst forth into human awareness in the early years of this century. The name of Einstein figures preeminently in the public conception of this field. It was, indeed, Albert Einstein who, in 1905, first clearly enunciated the profound principles which underlie this new realm of physical endeavor. But there were others before him, most notably Hendrik Antoon Lorentz and Henri Poincaré, who had already appreciated most of the basics of the (then) new physics. Moreover, the great scientists Galileo Galilei and Isaac Newton, centuries before Einstein, had already pointed out that in the dynamical theories that they themselves were developing, the physics as perceived by an observer in uniform motion would be identical with that perceived by an observer at rest. The key problem with this had arisen only later, with James Clerk Maxwell's discovery, as published in 1865, of the equations that govern the electric and magnetic fields, and which also control the propagation of light. The implication seemed to be that the relativity principle of Galileo and Newton could no longer hold true; for the speed of light must, by Maxwell's equations, have a definite speed of propagation. Accordingly, an observer at rest is distinguished from those in motion by the fact that only to an observer at rest does the light speed appear to be the same in all

Introduction

directions. The relativity principle of Lorentz, Poincaré, and Einstein differs from that of Galileo and Newton, but it has this same implication: the physics as perceived by an observer in uniform motion is indeed identical with that perceived by an observer at rest.

Yet, in the new relativity, Maxwell's equations *are* consistent with this principle, and the speed of light is measured to have a definite fixed value in every direction, no matter in what direction or with what speed the observer might be moving. How is this magic achieved so that these apparently hopelessly incompatible requirements are reconciled? I shall leave it to Feynman to explain—in his own inimitable fashion.

Relativity is perhaps the first place where the physical power of the mathematical idea of *symmetry* begins to be felt. Symmetry is a familiar idea, but it is less familiar to people how such an idea can be applied in accordance with a set of mathematical expressions. But it is just such a thing that is needed in order to implement the principles of special relativity in a system of equations. In order to be consistent with the relativity principle, whereby physics "looks the same" to an observer in uniform motion as to an observer at rest, there must be a "symmetry transformation" which translates one observer's measured quantities into those of the other. It is a symmetry because the physical laws appear the same to each observer, and "symmetry" after all, asserts that something has the same appearance from two distinct points of view. Feynman's approach to abstract matters of this nature is very down to earth, and he is able to convey the ideas in a way that is accessible to people with no particular mathematical experience or aptitude for abstract thinking.

Whereas relativity pointed the way to additional symmetries that had not been perceived before, some of the more modern developments in physics have shown that certain symmetries, previously thought to be universal, are in fact subtly violated. It came as one of the most profound shocks to the physical community in 1957, as the work of Lee, Yang, and Wu showed, that in certain basic physical processes, the laws satisfied by a physical system are not the same as those satisfied by the mirror reflection of that system. In fact, Feynman had a hand in the development of the physical theory which is able to accommodate this asymmetry. His account here is, accordingly, a dramatic one, as deeper and deeper mysteries of nature gradually unfold.

As physics develops, there are mathematical formalisms that develop with it, and which are needed in order to express the new physical laws. When the mathematical tools are skillfully tuned to their appropriate tasks, they can make the physics seem much simpler than otherwise. The ideas of vector calculus are a case in point. The vector calculus of three dimensions was originally developed to handle the physics of ordinary space, and it provides an invaluable piece of machinery for the expression of physical laws, such as those of Newton, where there is no physically preferred direction in space. To put this another way, the physical laws have a *symmetry* under ordinary rotations in space. Feynman brings home the power of the vector notation and the underlying ideas for expressing such laws.

Relativity theory, however, tells us that *time* should also be brought under the compass of these symmetry transformations, so a *four*-dimensional vector calculus is needed. This calculus is also introduced to us here by Feynman, as it provides the way of understanding how not only time and space must be considered as different aspects of the same four-dimensional structure, but the same is true of energy and momentum in the relativistic scheme.

The idea that the history of the universe should be viewed, physically, as a *four*-dimensional space-time, rather than as a three-dimensional space evolving with time is indeed fundamental to modern physics. It is an idea whose significance is not easy to grasp. Indeed, Einstein himself was not sympathetic to this idea when he first encountered it. The idea of space-time was not, in fact, Einstein's, although, in the popular imagination it is frequently attributed to him. It was the Russian/German geometer Hermann Minkowski, who had been a teacher of Einstein's at the Zurich

Introduction

Polytechnic, who first put forward the idea of four-dimensional space-time in 1908, a few years after Poincaré and Einstein had formulated special relativity theory. In a famous lecture, Minkowski asserted: "Henceforth space by itself, and time by itself, are doomed to fade away into mere shadows, and only a kind of unity between the two will preserve an independent reality."*

Feynman's most influential scientific discoveries, the ones that I have referred to above, stemmed from his own *space-time* approach to quantum mechanics. There is thus no question about the importance of space-time to Feynman's work and to modern physics generally. It is not surprising, therefore, that Feynman is forceful in his promotion of space-time ideas, stressing their physical significance. Relativity is not airy-fairy philosophy, nor is space-time mere mathematical formalism. It is a foundational ingredient of the very universe in which we live.

When Einstein became accustomed to the idea of space-time, he took it completely into his way of thinking. It became an essential part of his extension of special relativity—the relativity theory I have been referring to above that Lorentz, Poincaré, and Einstein introduced—to what is known as *general* relativity. In Einstein's general relativity, the space-time becomes *curved,* and it is able to incorporate the phenomenon of *gravity* into this curvature. Clearly, this is a difficult idea to grasp, and in Feynman's final lecture in this collection, he makes no attempt to describe the full mathematical machinery that is needed for the complete formulation of Einstein's theory. Yet he gives a powerfully dramatic description, with insightful use of intriguing analogies, in order to get the essential ideas across.

In all his lectures, Feynman made particular efforts to preserve accuracy in his descriptions, almost always qualifying what he says when there was any danger that his simplifications or analogies

---

* The Minkowski quote is from the Dover reprint of seminal publications on relativity *The Principle of Relativity* by Einstein, Lorentz, Weyl, and Minkowski (originally Methuen and Co., 1923).

Introduction

might be misleading or lead to erroneous conclusions. I felt, however, that his simplified account of the Einstein field equation of general relativity did need a qualification that he did not quite give. For in Einstein's theory, the "active" mass which is the source of gravity is not simply the same as the energy (according to Einstein's $E=mc^2$); instead, this source is the energy density *plus the sum of the pressures,* and it is this that is the source of gravity's inward accelerations. With this additional qualification, Feynman's account is superb, and provides an excellent introduction to this most beautiful and self-contained of physical theories.

While Feynman's lectures are unashamedly aimed at those who have aspirations to become physicists—whether professionally or in spirit only—they are undoubtedly accessible also to those with no such aspirations. Feynman strongly believed (and I agree with him) in the importance of conveying an understanding of our universe—according to the perceived basic principles of modern physics—far more widely than can be achieved merely by the teaching provided in physics courses. Even late in his life, when taking part in the investigations of the *Challenger* disaster, he took great pains to show, on national television, that the source of the disaster was something that could be appreciated at an ordinary level, and he performed a simple but convincing experiment on camera showing the brittleness of the shuttle's O-rings in cold conditions.

He was a showman, certainly, sometimes even a clown; but his overriding purpose was always serious. And what more serious purpose can there be than the understanding of the nature of our universe at its deepest levels? At conveying this understanding, Richard Feynman was supreme.

December 1996   ROGER PENROSE

Special Preface
(from *Lectures on Physics*)

Toward the end of his life, Richard Feynman's fame had transcended the confines of the scientific community. His exploits as a member of the commission investigating the space shuttle Challenger disaster gave him widespread exposure; similarly, a best-selling book about his picaresque adventures made him a folk hero almost of the proportions of Albert Einstein. But back in 1961, even before his Nobel Prize increased his visibility to the general public, Feynman was more than merely famous among members of the scientific community—he was legendary. Undoubtedly, the extraordinary power of his teaching helped spread and enrich the legend of Richard Feynman.

He was a truly great teacher, perhaps the greatest of his era and ours. For Feynman, the lecture hall was a theater, and the lecturer a performer, responsible for providing drama and fireworks as well as facts and figures. He would prowl about the front of a classroom, arms waving, "the impossible combination of theoretical physicist and circus barker, all body motion and sound effects," wrote *The New York Times*. Whether he addressed an audience of students, colleagues, or the general public, for those lucky enough to see Feynman lecture in person, the experience was usually unconventional and always unforgettable, like the man himself.

He was the master of high drama, adept at riveting the attention of every lecture-hall audience. Many years ago, he taught a course in Advanced Quantum Mechanics, a large class comprised of a few registered graduate students and most of the Caltech physics faculty. During one lecture, Feynman started explaining how to represent certain complicated integrals diagrammatically: time on this axis, space on that axis, wiggly line for this straight line, etc. Having described what is known to the world of physics as a Feynman diagram, he turned around to face the class, grinning wickedly. "And this is called *THE* diagram!" Feynman had reached the denouement, and the lecture hall erupted with spontaneous applause.

For many years after the lectures that make up this book were given, Feynman was an occasional guest lecturer for Caltech's freshman physics course. Naturally, his appearances had to be kept secret so there would be room left in the hall for the registered students. At one such lecture the subject was curved-space time, and Feynman was characteristically brilliant. But the unforgettable moment came at the beginning of the lecture. The supernova of 1987 has just been discovered, and Feynman was very excited about it. He said, "Tycho Brahe had his supernova, and Kepler had his. Then there weren't any for 400 years. But now I have mine." The class fell silent, and Feynman continued on. "There are 10^{11} stars in the galaxy. That used to be a *huge* number. But it's only a hundred billion. It's less than the national deficit! We used to call them astronomical numbers. Now we should call them economical numbers." The class dissolved in laughter, and Feynman, having captured his audience, went on with his lecture.

Showmanship aside, Feynman's pedagogical technique was simple. A summation of his teaching philosophy was found among his papers in the Caltech archives, in a note he had scribbled to himself while in Brazil in 1952:

"First figure out why you want the students to learn the subject and what you want them to know, and the method will result more or less by common sense."

What came to Feynman by "common sense" were often brilliant twists that perfectly captured the essence of his point. Once, during a public lecture, he was trying to explain why one must not verify an idea using the same data that suggested the idea in the first place. Seeming to wander off the subject, Feynman began talking about license plates. "You know, the most amazing thing happened to me tonight. I was coming here, on the way to the lecture, and I came in through the parking lot. And you won't believe what happened. I saw a car with the license plate ARW 357. Can you imagine? Of all the millions of license plates in the state, what was the chance that I would see that particular one tonight? Amazing!" A point that even many scientists fail to grasp was made clear through Feynman's remarkable "common sense."

In 35 years at Caltech (from 1952 to 1987), Feynman was listed as teacher of record for 34 courses. Twenty-five of them were advanced graduate courses, strictly limited to graduate students, unless undergraduates asked permission to take them (they often did, and permission was nearly always granted). The rest were mainly introductory graduate courses. Only once did Feynman teach courses purely for undergraduates, and that was the celebrated occasion in the academic years 1961 to 1962 and 1962 to 1963, with a brief reprise in 1964, when he gave the lectures that were to become *The Feynman Lectures on Physics*.

At the time there was a consensus at Caltech that freshman and sophomore students were getting turned off rather than spurred on by their two years of compulsory physics. To remedy the situation, Feynman was asked to design a series of lectures to be given to the students over the course of two years, first to freshmen, and then to the same class as sophomores. When he agreed, it was immediately decided that the lectures should be transcribed for publication. That job turned out to be far more difficult than anyone had imagined. Turning out publishable books required a tremendous amount of work on the part of his colleagues, as well as Feynman himself, who did the final editing of every chapter.

And the nuts and bolts of running a course had to be addressed. This task was greatly complicated by the fact that Feynman had only a vague outline of what he wanted to cover. This meant that no one knew what Feynman would say until he stood in front of a lecture hall filled with students and said it. The Caltech professors who assisted him would then scramble as best they could to handle mundane details, such as making up homework problems.

Why did Feynman devote more than two years to revolutionizing the way beginning physics was taught? One can only speculate, but there were probably three basic reasons. One is that he loved to have an audience, and this gave him a bigger theater than he usually had in graduate courses. The second was that he genuinely cared about students, and he simply thought that teaching freshmen was an important thing to do. The third and perhaps most important reason was the sheer challenge of reformulating physics, as he understood it, so that it could be presented to young students. This was his specialty, and was the standard by which he measured whether something was really understood. Feynman was once asked by a Caltech faculty member to explain why spin 1/2 particles obey Fermi-Dirac statistics. He gauged his audience perfectly and said, "I'll prepare a freshman lecture on it." But a few days later he returned and said, "You know, I couldn't do it. I couldn't reduce it to the freshman level. That means we really don't understand it."

This specialty of reducing deep ideas to simple, understandable terms is evident throughout *The Feynman Lectures on Physics*, but nowhere more so than in his treatment of quantum mechanics. To aficionados, what he has done is clear. He has presented, to beginning students, the path integral method, the technique of his own devising that allowed him to solve some of the most profound problems in physics. His own work using path integrals, among other achievements, led to the 1965 Nobel Prize that he shared with Julian Schwinger and Sin-Itero Tomanaga.

Through the distant veil of memory, many of the students and faculty attending the lectures have said that having two years of physics with Feynman was the experience of a lifetime. But that's not how it seemed at the time. Many of the students dreaded the class, and as the course wore on, attendance by the registered students started dropping alarmingly. But at the same time, more

Special Preface

and more faculty and graduate students started attending. The room stayed full, and Feynman may never have known he was losing some of his intended audience. But even in Feynman's view, his pedagogical endeavor did not succeed. He wrote in the 1963 preface to the *Lectures*: "I don't think I did very well by the students." Rereading the books, one sometimes seems to catch Feynman looking over his shoulder, not at his young audience, but directly at his colleagues, saying, "Look at that! Look how I finessed that point! Wasn't that clever?" But even when he thought he was explaining things lucidly to freshmen or sophomores, it was not really they who were able to benefit most from what he was doing. It was his peers—scientists, physicists, and professors—who would be the main beneficiaries of his magnificent achievement, which was nothing less than to see physics through the fresh and dynamic perspective of Richard Feynman.

Feynman was more than a great teacher. His gift was that he was an extraordinary teacher of teachers. If the purpose in giving *The Feynman Lectures on Physics* was to prepare a roomful of undergraduate students to solve examination problems in physics, he cannot be said to have succeeded particularly well. Moreover, if the intent was for the books to serve as introductory college textbooks, he cannot be said to have achieved his goal. Nevertheless, the books have been translated into ten foreign languages and are available in four bilingual editions. Feynman himself believed that his most important contribution to physics would not be QED, or the theory of superfluid helium, or polarons, or partons. His foremost contribution would be the three red books of *The Feynman Lectures on Physics*. That belief justifies this commemorative issue of these celebrated books.

DAVID L. GOODSTEIN
GERRY NEUGEBAUER
April 1989   California Institute of Technology

Feynman's Preface
(from *Lectures on Physics*)

These are the lectures in physics that I gave last year and the year before to the freshman and sophomore classes at Caltech. The lectures are, of course, not verbatim—they have been edited, sometimes extensively and sometimes less so. The lectures form only part of the complete course. The whole group of 180 students gathered in a big lecture room twice a week to hear these lectures and then they broke up into small groups of 15 to 20 students in recitation sections under the guidance of a teaching assistant. In addition, there was a laboratory session once a week.

The special problem we tried to get at with these lectures was to maintain the interest of the very enthusiastic and rather smart students coming out of the high schools and into Caltech. They have heard a lot about how interesting and exciting physics is—the theory of relativity, quantum mechanics, and other modern ideas. By the end of two years of our previous course, many would be very discouraged because there were really very few grand, new, modern ideas presented to them. They were made to study inclined planes, electrostatics, and so forth, and after two years it was quite stultifying. The problem was whether or not we could make a course which would save the more advanced and excited student by maintaining his enthusiasm.

The lectures here are not in any way meant to be a survey course, but are very serious. I thought to address them to the most intelligent in the class and to make sure, if possible, that even the most intelligent student was unable to completely encompass everything that was in the lectures—by putting in suggestions of applications of the ideas and concepts in various directions outside the main line of attack. For this reason, though, I tried very hard to make all the statements as accurate as possible, to point out in every case where the equations and ideas fitted into the body of physics, and how—when they learned more—things would be modified. I also felt that for such students it is important to indicate what it is that they should—if they are sufficiently clever—be able to understand by deduction from what has been said before, and what is being put in as something new. When new ideas came in, I would try either to deduce them if they were deducible, or to explain that it *was* a new idea which hadn't any basis in terms of things they had already learned and what was not supposed to be provable—but was just added in.

At the start of these lectures, I assumed that the students knew something when they came out of high school—such things as geometrical optics, simple chemistry ideas, and so on. I also didn't see that there was any reason to make the lectures in a definite order, in the sense that I would not be allowed to mention something until I was ready to discuss it in detail. There was a great deal of mention of things to come, without complete discussions. These more complete discussions would come later when the preparation became more advanced. Examples are the discussions of inductance, and of energy levels, which are at first brought in in a very qualitative way and are later developed more completely.

At the same time that I was aiming at the more active student, I also wanted to take care of the fellow for whom the extra fireworks and side applications are merely disquieting and who cannot be expected to learn most of the material in the lecture at all. For such a student, I wanted there to be at least a central core or backbone of material which he *could* get. Even if he didn't understand everything in a lecture, I hoped he wouldn't get nervous. I didn't expect him to understand everything, but only the central and most direct features. It takes, of course, a certain intelligence on his part to see

Feynman's Preface

which are the central theorems and central ideas, and which are the more advanced side issues and applications which he may understand only in later years.

In giving these lectures there was one serious difficulty: in the way the course was given, there wasn't any feedback from the students to the lecturer to indicate how well the lectures were going over. This is indeed a very serious difficulty, and I don't know how good the lectures really are. The whole thing was essentially an experiment. And if I did it again I wouldn't do it the same way—I hope I *don't* have to do it again! I think, though, that things worked out—so far as the physics is concerned—quite satisfactorily in the first year.

In the second year I was not so satisfied. In the first part of the course, dealing with electricity and magnetism, I couldn't think of any really unique or different way of doing it—of any way that would be particularly more exciting than the usual way of presenting it. So I don't think I did very much in the lectures on electricity and magnetism. At the end of the second year I had originally intended to go on, after the electricity and magnetism, by giving some more lectures on the properties of materials, but mainly to take up things like fundamental modes, solutions of the diffusion equation, vibrating systems, orthogonal functions, . . . developing the first stages of what are usually called "the mathematical methods of physics." In retrospect, I think that if I were doing it again I would go back to that original idea. But since it was not planned that I would be giving these lectures again, it was suggested that it might be a good idea to try to give an introduction to the quantum mechanics—what you will find in Volume III.

It is perfectly clear that students who will major in physics can wait until their third year for quantum mechanics. On the other hand, the argument was made that many of the students in our course study physics as a background for their primary interest in other fields. And the usual way of dealing with quantum mechanics makes that subject almost unavailable for the great majority of students because they have to take so long to learn it. Yet, in its real applications—especially in its more complex applications, such as in electrical engineering and chemistry—the full machinery of the differential equation approach is not actually used. So I tried to describe the principles of quantum mechanics in a way which wouldn't require that one first know the mathematics of partial differential equations. Even for a physicist I think that is an interesting thing to try to do—to present quantum mechanics in this reverse fashion—for several reasons which may be apparent in the lectures themselves. However, I think that the experiment in the quantum mechanics part was not completely successful—in large part because I really did not have enough time at the end (I should, for instance, have had three or four more lectures in order to deal more completely with such matters as energy bands and the spatial dependence of amplitudes). Also, I had never presented the subject this way before, so the lack of feedback was particularly serious. I now believe the quantum mechanics should be given at a later time. Maybe I'll have a chance to do it again someday. Then I'll do it right.

The reason there are no lectures on how to solve problems is because there were recitation sections. Although I did put in three lectures in the first year on how to solve problems, they are not included here. Also there was a lecture on inertial guidance which certainly belongs after the lecture on rotating systems, but which was, unfortunately, omitted. The fifth and sixth lectures are actually due to Matthew Sands, as I was out of town.

The question, of course, is how well this experiment has succeeded. My own point of view—which, however, does not seem to be shared by most of the people who worked with the students—is pessimistic. I don't think I did very well by the students. When I look at the way the majority of the students handled the problems on the examinations, I think that the system is a failure. Of course, my friends point out to me that there were one or two dozen students who—very surprisingly—understood almost everything in all of the lectures, and who were quite active in working with the material and worrying about the many points in an excited and interested way. These people have now, I believe, a first-rate back-

Feynman's Preface

ground in physics—and they are, after all, the ones I was trying to get at. But then, "The power of instruction is seldom of much efficacy except in those happy dispositions where it is almost super-fluous." (Gibbon)

Still, I didn't want to leave any student completely behind, as perhaps I did. I think one way we could help the students more would be by putting more hard work into developing a set of problems which would elucidate some of the ideas in the lectures. Problems give a good opportunity to fill out the material of the lectures and make more realistic, more complete, and more settled in the mind the ideas that have been exposed.

I think, however, that there isn't any solution to this problem of education other than to realize that the best teaching can be done only when there is a direct individual relationship between a student and a good teacher—a situation in which the student discusses the ideas, thinks about the things, and talks about the things. It's impossible to learn very much by simply sitting in a lecture, or even by simply doing problems that are assigned. But in our modern times we have so many students to teach that we have to try to find some substitute for the ideal. Perhaps my lectures can make some contribution. Perhaps in some small place where there are individual teachers and students, they may get some inspiration or some ideas from the lectures. Perhaps they will have fun thinking them through—or going on to develop some of the ideas further.

June 1963   RICHARD P. FEYNMAN

SIX
NOT-SO-EASY
PIECES

One

VECTORS

1-1 Symmetry in physics

▶ In this chapter we introduce a subject that is technically known in physics as *symmetry in physical law*. The word "symmetry" is used here with a special meaning, and therefore needs to be defined. When is a thing symmetrical—how can we define it? When we have a picture that is symmetrical, one side is somehow the same as the other side. Professor Hermann Weyl has given this definition of symmetry: a thing is symmetrical if one can subject it to a certain operation and it appears exactly the same after the operation. For instance, if we look at a vase that is left-and-right symmetrical, then turn it 180° around the vertical axis, it looks the same. We shall adopt the definition of symmetry in Weyl's more general form, and in that form we shall discuss symmetry of physical laws.

Suppose we build a complex machine in a certain place, with a lot of complicated interactions, and balls bouncing around with forces between them, and so on. Now suppose we build exactly the same kind of equipment at some other place, matching part by part, with the same dimensions and the same orientation, everything the same only displaced laterally by some distance. Then, if we start the two machines in the same initial circumstances, in exact correspondence, we ask: will one machine behave exactly the same as the other? Will it follow all the motions in exact parallelism? Of course the answer may well be *no*, because if we choose the wrong place for

1

our machine it might be inside a wall and interferences from the wall would make the machine not work.

All of our ideas in physics require a certain amount of common sense in their application; they are not purely mathematical or abstract ideas. We have to understand what we mean when we say that the phenomena are the same when we move the apparatus to a new position. We mean that we move everything that we believe is relevant; if the phenomenon is not the same, we suggest that something relevant has not been moved, and we proceed to look for it. If we never find it, then we claim that the laws of physics do not have this symmetry. On the other hand, we may find it—we expect to find it—if the laws of physics do have this symmetry; looking around, we may discover, for instance, that the wall is pushing on the apparatus. The basic question is, if we define things well enough, if all the essential forces are included inside the apparatus, if all the relevant parts are moved from one place to another, will the laws be the same? Will the machinery work the same way?

It is clear that what we want to do is to move all the equipment and *essential* influences, but not *everything* in the world—planets, stars, and all—for if we do that, we have the same phenomenon again for the trivial reason that we are right back where we started. No, we cannot move *everything*. But it turns out in practice that with a certain amount of intelligence about what to move, the machinery will work. In other words, if we do not go inside a wall, if we know the origin of the outside forces, and arrange that those are moved too, then the machinery *will* work the same in one location as in another.

1-2 Translations

We shall limit our analysis to just mechanics, for which we now have sufficient knowledge. In previous chapters we have seen that the laws of mechanics can be summarized by a set of three equations for each particle:

$$m(d^2x/dt^2) = F_x, \quad m(d^2y/dt^2) = F_y, \quad m(d^2z/dt^2) = F_z. \qquad (1.1)$$

Vectors

Now this means that there exists a way to *measure x, y,* and *z* on three perpendicular axes, and the forces along those directions, such that these laws are true. These must be measured from some origin, but *where do we put the origin?* All that Newton would tell us at first is that there *is* some place that we can measure from, perhaps the center of the universe, such that these laws are correct. But we can show immediately that we can never find the center, because if we use some other origin it would make no difference. In other words, suppose that there are two people—Joe, who has an origin in one place, and Moe, who has a parallel system whose origin is somewhere else (Figure 1-1). Now when Joe measures the location of the point in space, he finds it at *x, y,* and *z* (we shall usually leave *z* out because it is too confusing to draw in a picture). Moe, on the other hand, when measuring the same point, will obtain a different *x* (in order to distinguish it, we will call it *x'*), and in principle a different *y*, although in our example they are numerically equal. So we have

$$x' = x - a, \quad y' = y, \quad z' = z. \tag{1.2}$$

Now in order to complete our analysis we must know what Moe would obtain for the forces. The force is supposed to act along some line, and by the force in the *x*-direction we mean the part of the total which is in the *x*-direction, which is the magnitude of the force times this cosine of its angle with the *x*-axis. Now we see that

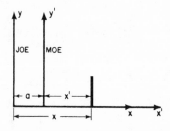

Figure 1-1. Two parallel coordinate systems.

Moe would use exactly the same projection as Joe would use, so we have a set of equations

$$F_{x'} = F_x, \quad F_{y'} = F_y, \quad F_{z'} = F_z. \tag{1.3}$$

These would be the relationships between quantities as seen by Joe and Moe.

The question is, if Joe knows Newton's laws, and if Moe tries to write down Newton's laws, will they also be correct for him? Does it make any difference from which origin we measure the points? In other words, assuming that equations (1.1) are true, and the Eqs. (1.2) and (1.3) give the relationship of the measurements, is it or is it not true that

(a) $m(d^2x'/dt^2) = F_{x'},$

(b) $m(d^2y'/dt^2) = F_{y'},$ \hspace{1cm} (1.4)

(c) $m(d^2z'/dt^2) = F_{z'}?$

In order to test these equations we shall differentiate the formula for x' twice. First of all

$$\frac{dx'}{dt} = \frac{d}{dt}(x - a) = \frac{dx}{dt} - \frac{da}{dt}.$$

Now we shall assume that Moe's origin is fixed (not moving) relative to Joe's; therefore a is a constant and $da/dt = 0$, so we find that

$$dx'/dt = dx/dt$$

and therefore

$$d^2x'/dt^2 = d^2x/dt^2;$$

therefore we know that Eq. (1.4a) becomes

$$m(d^2x/dt^2) = F_{x'}.$$

(We also suppose that the masses measured by Joe and Moe are equal.) Thus the acceleration times the mass is the same as the other fellow's. We have also found the formula for $F_{x'}$, for, substituting from Eq. (1.1), we find that

$$F_{x'} = F_x.$$

Vectors

Therefore the laws as seen by Moe appear the same; he can write Newton's laws too, with different coordinates, and they will still be right. That means that there is no unique way to define the origin of the world, because the laws will appear the same, from whatever position they are observed.

This is also true: if there is a piece of equipment in one place with a certain kind of machinery in it, the same equipment in another place will behave in the same way. Why? Because one machine, when analyzed by Moe, has exactly the same equations as the other one, analyzed by Joe. Since the *equations* are the same, the *phenomena* appear the same. So the proof that an apparatus in a new position behaves the same as it did in the old position is the same as the proof that the equations when displaced in space reproduce themselves. Therefore we say that *the laws of physics are symmetrical for translational displacements*, symmetrical in the sense that the laws do not change when we make a translation of our coordinates. Of course it is quite obvious intuitively that this is true, but it is interesting and entertaining to discuss the mathematics of it.

1-3 Rotations

The above is the first of a series of ever more complicated propositions concerning the symmetry of a physical law. The next proposition is that it should make no difference in which *direction* we choose the axes. In other words, if we build a piece of equipment in some place and watch it operate, and nearby we build the same kind of apparatus but put it up on an angle, will it operate in the same way? Obviously it will not if it is a grandfather clock, for example! If a pendulum clock stands upright, it works fine, but if it is tilted the pendulum falls against the side of the case and nothing happens. The theorem is then false in the case of the pendulum clock, unless we include the earth, which is pulling on the pendulum. Therefore we can make a prediction about pendulum clocks if we believe in the symmetry of physical law for rotation: something else is involved in the operation of a pendulum clock besides the machinery of the clock, something outside it that we should look for. We may also predict that pendulum clocks will not work the same way when located in different places relative to this mysterious source of asymmetry, perhaps the earth. Indeed, we know that a pendulum clock up in an artificial satellite, for example, would not tick either, because there is no effective force, and on Mars it would go at a different rate. Pendulum clocks *do* involve something more than just the machinery inside, they involve something on the outside. Once we recognize this factor, we see that we must turn the earth along with the apparatus. Of course we do not have to worry about that, it is easy to do; one simply waits a moment or two and the earth turns; then the pendulum clock ticks again in the new position the same as it did before. While we are rotating in space our angles are always changing, absolutely; this change does not seem to bother us very much, for in the new position we seem to be in the same condition as in the old. This has a certain tendency to confuse one, because it is true that in the new turned position the laws are the same as in the unturned position, but it is *not* true that *as we turn* a thing it follows the same laws as it does when we are not turning it. If we perform sufficiently delicate experiments, we can tell that the earth *is rotating*, but not that it *had rotated*. In other words, we cannot locate its angular position, but we can tell that it is changing.

Now we may discuss the effects of angular orientation upon physical laws. Let us find out whether the same game with Joe and Moe works again. This time, to avoid needless complication, we shall suppose that Joe and Moe use the same origin (we have already shown that the axes can be moved by translation to another place). Assume that Moe's axes have rotated relative to Joe's by an angle θ. The two coordinate systems are shown in Figure 1-2, which is restricted to two dimensions. Consider any point P having coordinates (x, y) in Joe's system and (x', y') in Moe's system. We shall begin, as in the previous case, by expressing the coordinates x' and y' in terms of x, y, and θ. To do so, we first drop perpendiculars from P to all four axes and draw AB perpendicular to PQ. Inspection of the figure shows that x' can be written as the sum of two

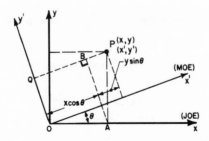

Figure 1-2. Two coordinate systems having different angular orientations.

lengths along the x'-axis, and y' as the difference of two lengths along AB. All these lengths are expressed in terms of x, y, and θ in equations (1.5), to which we have added an equation for the third dimension.

$$
\begin{aligned}
x' &= x \cos \theta + y \sin \theta, \\
y' &= y \cos \theta - x \sin \theta, \\
z' &= z.
\end{aligned} \tag{1.5}
$$

The next step is to analyze the relationship of forces as seen by the two observers, following the same general method as before. Let us assume that a force F, which has already been analyzed as having components F_x and F_y (as seen by Joe), is acting on a particle of mass m, located at point P in Figure 1-2. For simplicity, let us move both sets of axes so that the origin is at P, as shown in Figure 1-3. Moe sees the components of F along his axes as $F_{x'}$ and $F_{y'}$. F_x has components along both the x'- and y'-axes, and F_y likewise has components along both these axes. To express $F_{x'}$ in terms of F_x and F_y, we sum these components along the x'-axis, and in a like manner we can express $F_{y'}$ in terms of F_x and F_y. The results are

$$
\begin{aligned}
F_{x'} &= F_x \cos \theta + F_y \sin \theta, \\
F_{y'} &= F_y \cos \theta - F_x \sin \theta, \\
F_{z'} &= F_z.
\end{aligned} \tag{1.6}
$$

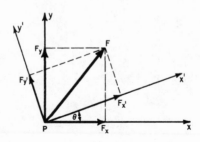

Figure 1-3. Components of a force in the two systems.

It is interesting to note an accident of sorts, which is of extreme importance: the formulas (1.5) and (1.6), for coordinates of P and components of F, respectively, *are of identical form.*

As before, Newton's laws are assumed to be true in Joe's system, and are expressed by equations (1.1). The question, again, is whether Moe can apply Newton's laws—will the results be correct for his system of rotated axes? In other words, if we assume that Eqs. (1.5) and (1.6) give the relationship of the measurements, is it true or not true that

$$m(d^2x'/dt^2) = F_{x'},$$
$$m(d^2y'/dt^2) = F_{y'}, \quad\quad\quad (1.7)$$
$$m(d^2z'/dt^2) = F_{z'}?$$

To test these equations, we calculate the left and right sides independently, and compare the results. To calculate the left sides, we multiply equations (1.5) by m, and differentiate twice with respect to time, assuming the angle θ to be constant. This gives

$$m(d^2x'/dt^2) = m(d^2x/dt^2) \cos\theta + m(d^2y/dt^2) \sin\theta,$$
$$m(d^2y'/dt^2) = m(d^2y/dt^2) \cos\theta - m(d^2x/dt^2) \sin\theta, \quad (1.8)$$
$$m(d^2z'/dt^2) = m(d^2z/dt^2).$$

We calculate the right sides of equations (1.7) by substituting equations (1.1) into equations (1.6). This gives

$$F_{x'} = m(d^2x/dt^2) \cos \theta + m(d^2y/dt^2) \sin \theta,$$
$$F_{y'} = m(d^2y/dt^2) \cos \theta - m(d^2x/dt^2) \sin \theta, \qquad (1.9)$$
$$F_{z'} = m(d^2z/dt^2).$$

Behold! The right sides of Eqs. (1.8) and (1.9) are identical, so we conclude that if Newton's laws are correct on one set of axes, they are also valid on any other set of axes. This result, which has now been established for both translation and rotation of axes, has certain consequences: first, no one can claim his particular axes are unique, but of course they can be more *convenient* for certain particular problems. For example, it is handy to have gravity along one axis, but this is not physically necessary. Second, it means that any piece of equipment which is completely self-contained, with all the force-generating equipment completely inside the apparatus, would work the same when turned at an angle.

1-4 Vectors

Not only Newton's laws, but also the other laws of physics, so far as we know today, have the two properties which we call invariance (or symmetry) under translation of axes and rotation of axes. These properties are so important that a mathematical technique has been developed to take advantage of them in writing and using physical laws.

The foregoing analysis involved considerable tedious mathematical work. To reduce the details to a minimum in the analysis of such questions, a very powerful mathematical machinery has been devised. This system, called *vector analysis*, supplies the title of this chapter; strictly speaking, however, this is a chapter on the symmetry of physical laws. By the methods of the preceding analysis we were able to do everything required for obtaining the results that we sought, but in practice we should like to do things more easily and rapidly, so we employ the vector technique.

We begin by noting some characteristics of two kinds of quantities that are important in physics. (Actually there are more than two, but let us start out with two.) One of them, like the number of potatoes in a sack, we call an ordinary quantity, or an undirected quantity, or a *scalar*. Temperature is an example of such a quantity. Other quantities that are important in physics do have direction, for instance velocity: we have to keep track of which way a body is going, not just its speed. Momentum and force also have direction, as does displacement: when someone steps from one place to another in space, we can keep track of how far he went, but if we wish also to know *where* he went, we have to specify a direction.

All quantities that have a direction, like a step in space, are called *vectors*.

A vector is three numbers. In order to represent a step in space, say from the origin to some particular point P whose location is (x, y, z), we really need three numbers, but we are going to invent a single mathematical symbol, $\mathbf{r}$, which is unlike any other mathematical symbols we have so far used.* It is *not* a single number, it represents *three* numbers: x, y, and z. It means three numbers, but not really only *those* three numbers, because if we were to use a different coordinate system, the three numbers would be changed to x', y', and z'. However, we want to keep our mathematics simple and so we are going to use the *same mark* to represent the three numbers (x, y, z) and the three numbers (x', y', z'). That is, we use the same mark to represent the first set of three numbers for one coordinate system, but the second set of three numbers if we are using the other coordinate system. This has the advantage that when we change the coordinate system, we do not have to change the letters of our equations. If we write an equation in terms of x, y, z, and then use another system, we have to change to x', y', z', but we shall just write $\mathbf{r}$, with the convention that it represents (x, y, z) if we use one set of axes, or (x', y', z') if we use another set of axes, and so on. The three numbers which describe the quantity in a given coordinate system are called the *components* of the vector in the

---

* In type, vectors are represented by boldface; in handwritten form an arrow is used: $\vec{r}$.

Vectors

direction of the coordinate axes of that system. That is, we use the same symbol for the three letters that correspond to the *same object, as seen from different axes*. The very fact that we can say "the same object" implies a physical intuition about the reality of a step in space, that is independent of the components in terms of which we measure it. So the symbol $\mathbf{r}$ will represent the same thing no matter how we turn the axes.

Now suppose there is another directed physical quantity, any other quantity, which also has three numbers associated with it, like force, and these three numbers change to three other numbers by a certain mathematical rule, if we change the axes. It must be the same rule that changes (x, y, z) into (x', y', z'). In other words, any physical quantity associated with three numbers which transform as do the components of a step in space is a vector. An equation like

$$\mathbf{F} = \mathbf{r}$$

would thus be true in *any* coordinate system if it were true in one. This equation, of course, stands for the three equations

$$F_x = x, \quad F_y = y, \quad F_z = z,$$

or, alternatively, for

$$F_{x'} = x', \quad F_{y'} = y', \quad F_{z'} = z'.$$

The fact that a physical relationship can be expressed as a vector equation assures us the relationship is unchanged by a mere rotation of the coordinate system. That is the reason why vectors are so useful in physics.

Now let us examine some of the properties of vectors. As examples of vectors we may mention velocity, momentum, force, and acceleration. For many purposes it is convenient to represent a vector quantity by an arrow that indicates the direction in which it is acting. Why can we represent force, say, by an arrow? Because it has the same mathematical transformation properties as a "step in space." We thus represent it in a diagram as if it were a step, using a scale such that one unit of force, or one newton, corresponds to a certain convenient length. Once we have done this, all forces can be represented as lengths, because an equation like

$$\mathbf{F} = k\mathbf{r},$$

where k is some constant, is a perfectly legitimate equation. Thus we can always represent forces by lines, which is very convenient, because once we have drawn the line we no longer need the axes. Of course, we can quickly calculate the three components as they change upon turning the axes, because that is just a geometric problem.

1-5 Vector algebra

Now we must describe the laws, or rules, for combining vectors in various ways. The first such combination is the *addition* of two vectors: suppose that **a** is a vector which in some particular coordinate system has the three components (a_x, a_y, a_z), and that **b** is another vector which has the three components (b_x, b_y, b_z). Now let us invent three new numbers $(a_x + b_x, a_y + b_y, a_z + b_z)$. Do these form a vector? "Well," we might say, "they are three numbers, and every three numbers form a vector." No, *not* every three numbers form a vector! In order for it to be a vector, not only must there be three numbers, but these must be associated with a coordinate system in such a way that if we turn the coordinate system, the three numbers "revolve" on each other, get "mixed up" in each other, by the precise laws we have already described. So the question is, if we now rotate the coordinate system so that (a_x, a_y, a_z) become $(a_{x'}, a_{y'}, a_{z'})$ and (b_x, b_y, b_z) become $(b_{x'}, b_{y'}, b_{z'})$, what do $(a_x + b_x, a_y + b_y, a_z + b_z)$ become? Do they become $(a_{x'} + b_{x'}, a_{y'} + b_{y'}, a_{z'} + b_{z'})$ or not? The answer is, of course, yes, because the prototype transformations of Eq. (1.5) constitute what we call a *linear* transformation. If we apply those transformations to a_x and b_x to get $a_{x'} + b_{x'}$, we find that the transformed $a_x + b_x$ is indeed the same as $a_{x'} + b_{x'}$. When **a** and **b** are "added together" in this sense, they will form a vector which we may call **c**. We would write this as

Vectors

$$c = a + b.$$

Now **c** has the interesting property

$$c = b + a,$$

as we can immediately see from its components. Thus also,

$$a + (b + c) = (a + b) + c.$$

We can add vectors in any order.

What is the geometric significance of **a** + **b**? Suppose that **a** and **b** were represented by lines on a piece of paper, what would **c** look like? This is shown in Figure 1-4. We see that we can add the components of **b** to those of **a** most conveniently if we place the rectangle representing the components of **b** next to that representing the components of **a** in the manner indicated. Since **b** just "fits" into its rectangle, as does **a** into its rectangle, this is the same as putting the "tail" of **b** on the "head" of **a**, the arrow from the "tail" of **a** to the "head" of **b** being the vector **c**. Of course, if we added **a** to **b** the other way around, we would put the "tail" of **a** on the "head" of **b**, and by the geometrical properties of parallelograms we would get the same result for **c**. Note that vectors can be added in this way without reference to any coordinate axes.

Suppose we multiply a vector by a number α, what does this mean? We *define* it to mean a new vector whose components are

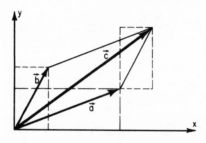

Figure 1-4. The addition of vectors.

αa_x, αa_y, and αa_z. We leave it as a problem for the student to prove that it *is* a vector.

Now let us consider vector subtraction. We may define subtraction in the same way as addition, but instead of adding, we subtract the components. Or we might define subtraction by defining a negative vector, $-\mathbf{b} = -1\mathbf{b}$, and then we would add the components. It comes to the same thing. The result is shown in Figure 1-5. This figure shows $\mathbf{d} = \mathbf{a} - \mathbf{b} = \mathbf{a} + (-\mathbf{b})$; we also note that the difference $\mathbf{a} - \mathbf{b}$ can be found very easily from $\mathbf{a}$ and $\mathbf{b}$ by using the equivalent relation $\mathbf{a} = \mathbf{b} + \mathbf{d}$. Thus the difference is even easier to find than the sum: we just draw the vector from $\mathbf{b}$ to $\mathbf{a}$, to get $\mathbf{a} - \mathbf{b}$!

Next we discuss velocity. Why is velocity a vector? If position is given by the three coordinates (x, y, z), what is the velocity? The velocity is given by dx/dt, dy/dt, and dz/dt. Is that a vector, or not? We can find out by differentiating the expressions in Eq. (1.5) to find out whether dx'/dt *transforms* in the right way. We see that the components dx/dt and dy/dt *do* transform according to the same law as x and y, and therefore the time derivative is a vector. So the velocity *is* a vector. We can write the velocity in an interesting way as

$$\mathbf{v} = d\mathbf{r}/dt.$$

What the velocity is, and why it is a vector, can also be understood more pictorially: How far does a particle move in a short time Δt?

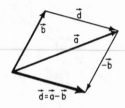

Figure 1-5. The subtraction of vectors.

Vectors

Answer: **Δr**, so if a particle is "here" at one instant and "there" at another instant, then the vector difference of the positions **Δr** = **r**$_2$ − **r**$_1$, which is in the direction of motion shown in Figure 1-6, divided by the time interval $\Delta t = t_2 - t_1$, is the "average velocity" vector.

In other words, by vector velocity we mean the limit, as Δt goes to 0, of the difference between the radius vectors at the time $t + \Delta t$ and the time t, divided by Δt:

$$\mathbf{v} = \lim_{\Delta t \to 0} (\Delta \mathbf{r}/\Delta t) = d\mathbf{r}/dt. \qquad (1.10)$$

Thus velocity is a vector because it is the difference of two vectors. It is also the right definition of velocity because its components are dx/dt, dy/dt, and dz/dt. In fact, we see from this argument that if we differentiate *any* vector with respect to time we produce a new vector. So we have several ways of producing new vectors: (1) multiply by a constant, (2) differentiate with respect to time, (3) add or subtract two vectors.

1-6 Newton's laws in vector notation

In order to write Newton's laws in vector form, we have to go just one step further, and define the acceleration vector. This is the time derivative of the velocity vector, and it is easy to demonstrate that its

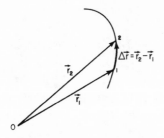

Figure 1-6. The displacement of a particle in a short time interval $\Delta t = t_2 - t_1$.

components are the second derivatives of x, y, and z with respect to t:

$$\mathbf{a} = \frac{d\mathbf{v}}{dt} = \left(\frac{d}{dt}\right)\left(\frac{d\mathbf{r}}{dt}\right) = \frac{d^2\mathbf{r}}{dt^2}, \tag{1.11}$$

$$a_x = \frac{dv_x}{dt} = \frac{d^2x}{dt^2}, \quad a_y = \frac{dv_y}{dt} = \frac{d^2y}{dt^2}, \quad a_z = \frac{dv_z}{dt} = \frac{d^2z}{dt^2}. \tag{1.12}$$

With this definition, then, Newton's laws can be written in this way:

$$m\mathbf{a} = \mathbf{F} \tag{1.13}$$

or

$$m(d^2\mathbf{r}/dt^2) = \mathbf{F}. \tag{1.14}$$

Now the problem of proving the invariance of Newton's laws under rotation of coordinates is this: prove that $\mathbf{a}$ is a vector; this we have just done. Prove that $\mathbf{F}$ is a vector; we *suppose* it is. So if force is a vector, then, since we know acceleration is a vector, Eq. (1.13) will look the same in any coordinate system. Writing it in a form which does not explicitly contain x's, y's, and z's has the advantage that from now on we need not write *three* laws every time we write Newton's equations or other laws of physics. We write what looks like *one* law, but really, of course, it is the three laws for any particular set of axes, because any vector equation involves the statement that *each of the components is equal.*

The fact that the acceleration is the rate of change of the vector velocity helps us to calculate the acceleration in some rather complicated circumstances. Suppose, for instance, that a particle is moving on some complicated curve (Figure 1-7) and that, at a given instant t, it had a certain velocity $\mathbf{v}_1$, but that when we go to another instant t_2 a little later, it has a different velocity $\mathbf{v}_2$. What is the acceleration? Answer: Acceleration is the difference in the velocity divided by the small time interval, so we need the difference of the two velocities. How do we get the difference of the velocities? To subtract two vectors, we put the vector across the ends of $\mathbf{v}_2$ and $\mathbf{v}_1$; that is, we

Vectors

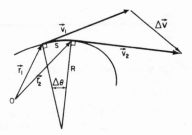

Figure 1-7. A curved trajectory.

draw Δ as the difference of the two vectors, right? *No!* That only works when the *tails* of the vectors are in the same place! It has no meaning if we move the vector somewhere else and then draw a line across, so watch out! We have to draw a new diagram to subtract the vectors. In Figure 1-8, $\mathbf{v}_1$ and $\mathbf{v}_2$ are both drawn parallel and equal to their counterparts in Figure 1-7, and now we can discuss the acceleration. Of course the acceleration is simply $\Delta\mathbf{v}/\Delta t$. It is interesting to note that we can compose the velocity difference out of two parts; we can think of acceleration as having *two components*, $\Delta\mathbf{v}_\parallel$ in the direction tangent to the path and $\Delta\mathbf{v}_\perp$ at right angles to the path, as indicated in Figure 1-8. The acceleration tangent to the path is, of course, just the change in the *length* of the vector, i.e., the change in the *speed v*:

$$a_\parallel = dv/dt. \tag{1.15}$$

The other component of acceleration, at right angles to the curve, is easy to calculate, using Figures 1-7 and 1-8. In the short time Δt let the change in angle between $\mathbf{v}_1$ and $\mathbf{v}_2$ be the small angle $\Delta\theta$. If

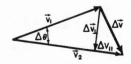

Figure 1-8. Diagram for calculating the acceleration.

the magnitude of the velocity is called v, then of course

$$\Delta v_\perp = v \, \Delta\theta$$

and the acceleration a will be

$$a_\perp = v \, (\Delta\theta/\Delta t).$$

Now we need to know $\Delta\theta/\Delta t$, which can be found this way: If, at the given moment, the curve is approximated as a circle of a certain radius R, then in a time Δt the distance s is, of course, $v \, \Delta t$, where v is the speed.

$$\Delta\theta = v(\Delta t/R), \quad \text{or} \quad \Delta\theta/\Delta t = v/R.$$

Therefore, we find

$$a = v^2/R, \tag{1.16}$$

as we have seen before.

1-7 Scalar product of vectors

Now let us examine a little further the properties of vectors. It is easy to see that the *length* of a step in space would be the same in any coordinate system. That is, if a particular step $\mathbf{r}$ is represented by x, y, z, in one coordinate system, and by x', y', z' in another coordinate system, surely the distance $r = |\mathbf{r}|$ would be the same in both. Now

$$r = \sqrt{x^2 + y^2 + z^2}$$

and also

$$r' = \sqrt{x'^2 + y'^2 + z'^2}.$$

So what we wish to verify is that these two quantities are equal. It is much more convenient not to bother to take the square root, so let us talk about the square of the distance; that is, let us find out whether

$$x^2 + y^2 + z^2 = x'^2 + y'^2 + z'^2. \tag{1.17}$$

It had better be—and if we substitute Eq. (1.5) we do indeed find that it is. So we see that there are other kinds of equations which are true for any two coordinate systems.

Vectors

Something new is involved. We can produce a new quantity, a function of x, y, and z, called a *scalar function*, a quantity which has no direction but which is the same in both systems. Out of a vector we can make a scalar. We have to find a general rule for that. It is clear what the rule is for the case just considered: add the squares of the components. Let us now define a new thing, which we call $\mathbf{a} \cdot \mathbf{a}$. This is not a vector, but a scalar; it is a number that is the same in all coordinate systems, and it is defined to be the sum of the squares of the three components of the vector:

$$\mathbf{a} \cdot \mathbf{a} = a_x^2 + a_y^2 + a_z^2. \tag{1.18}$$

Now you say, "But with what axes?" It does not depend on the axes, the answer is the same in *every* set of axes. So we have a new *kind* of quantity, a new *invariant* or *scalar* produced by one vector "squared." If we now define the following quantity for any two vectors $\mathbf{a}$ and $\mathbf{b}$:

$$\mathbf{a} \cdot \mathbf{b} = a_x b_x + a_y b_y + a_z b_z, \tag{1.19}$$

we find that this quantity, calculated in the primed and unprimed systems, also stays the same. To prove it we note that it is true of $\mathbf{a} \cdot \mathbf{a}$, $\mathbf{b} \cdot \mathbf{b}$, and $\mathbf{c} \cdot \mathbf{c}$, where $\mathbf{c} = \mathbf{a} + \mathbf{b}$. Therefore the sum of the squares $(a_x + b_x)^2 + (a_y + b_y)^2 + (a_z + b_z)^2$ will be invariant:

$$(a_x + b_x)^2 + (a_y + b_y)^2 + (a_z + b_z)^2 = $$
$$(a_{x'} + b_{x'})^2 + (a_{y'} + b_{y'})^2 + (a_{z'} + b_{z'})^2. \tag{1.20}$$

If both sides of this equation are expanded, there will be cross products of just the type appearing in Eq. (1.19), as well as the sums of squares of the components of $\mathbf{a}$ and $\mathbf{b}$. The invariance of terms of the form of Eq. (1.18) then leaves the cross product terms (1.19) invariant also.

The quantity $\mathbf{a} \cdot \mathbf{b}$ is called the *scalar product* of two vectors, $\mathbf{a}$ and $\mathbf{b}$, and it has many interesting and useful properties. For instance, it is easily proved that

$$\mathbf{a} \cdot (\mathbf{b} + \mathbf{c}) = \mathbf{a} \cdot \mathbf{b} + \mathbf{a} \cdot \mathbf{c}. \tag{1.21}$$

SIX NOT-SO-EASY PIECES

Also, there is a simple geometrical way to calculate $\mathbf{a} \cdot \mathbf{b}$, without having to calculate the components of $\mathbf{a}$ and $\mathbf{b}$: $\mathbf{a} \cdot \mathbf{b}$ is the product of the length of $\mathbf{a}$ and the length of $\mathbf{b}$ times the cosine of the angle between them. Why? Suppose that we choose a special coordinate system in which the x-axis lies along $\mathbf{a}$; in those circumstances, the only component of $\mathbf{a}$ that will be there is a_x, which is of course the whole length of $\mathbf{a}$. Thus Eq. (1.19) reduces to $\mathbf{a} \cdot \mathbf{b} = a_x b_x$ for this case, and this is the length of $\mathbf{a}$ times the component of $\mathbf{b}$ in the direction of $\mathbf{a}$, that is, $b \cos \theta$:

$$\mathbf{a} \cdot \mathbf{b} = ab \cos \theta.$$

Therefore, in that special coordinate system, we have proved that $\mathbf{a} \cdot \mathbf{b}$ is the length of $\mathbf{a}$ times the length of $\mathbf{b}$ times $\cos \theta$. But *if it is true in one coordinate system, it is true in all*, because $\mathbf{a} \cdot \mathbf{b}$ is independent of the coordinate system; that is our argument.

What good is the dot product? Are there any cases in physics where we need it? Yes, we need it all the time. For instance, in Chapter 4* the kinetic energy was called $\frac{1}{2}mv^2$, but if the object is moving in space it should be the velocity squared in the x-direction, the y-direction, and the z-direction, and so the formula for kinetic energy according to vector analysis is

$$\text{K.E.} = \tfrac{1}{2}m(\mathbf{v} \cdot \mathbf{v}) = \tfrac{1}{2}m(v_x^2 + v_y^2 + v_z^2). \tag{1.22}$$

Energy does not have direction. Momentum has direction; it is a vector, and it is the mass times the velocity vector.

Another example of a dot product is the work done by a force when something is pushed from one place to the other. We have not yet defined work, but it is equivalent to the energy change, the weights lifted, when a force $\mathbf{F}$ acts through a distance $\mathbf{s}$:

$$\text{Work} = \mathbf{F} \cdot \mathbf{s}. \tag{1.23}$$

It is sometimes very convenient to talk about the component of a vector in a certain direction (say the vertical direction because that is the direction of gravity). For such purposes, it is useful to invent

* of the original *Lectures on Physics*, vol. I.

what we call a *unit vector* in the direction that we want to study. By a unit vector we mean one whose dot product with itself is equal to unity. Let us call this unit vector $\mathbf{i}$; then $\mathbf{i} \cdot \mathbf{i} = 1$. Then, if we want the component of some vector in the direction of $\mathbf{i}$, we see that the dot product $\mathbf{a} \cdot \mathbf{i}$ will be $a \cos \theta$, i.e., the component of $\mathbf{a}$ in the direction of $\mathbf{i}$. This is a nice way to get the component; in fact, it permits us to get *all* the components and to write a rather amusing formula. Suppose that in a given system of coordinates, x, y, and z, we invent three vectors: $\mathbf{i}$, a unit vector in the direction x; $\mathbf{j}$, a unit vector in the direction y; and $\mathbf{k}$, a unit vector in the direction z. Note first that $\mathbf{i} \cdot \mathbf{i} = 1$. What is $\mathbf{i} \cdot \mathbf{j}$? When two vectors are at right angles, their dot product is zero. Thus

$$\begin{aligned} \mathbf{i} \cdot \mathbf{i} &= 1 \\ \mathbf{i} \cdot \mathbf{j} &= 0 \quad \mathbf{j} \cdot \mathbf{j} = 1 \\ \mathbf{i} \cdot \mathbf{k} &= 0 \quad \mathbf{j} \cdot \mathbf{k} = 0 \quad \mathbf{k} \cdot \mathbf{k} = 1 \end{aligned} \tag{1.24}$$

Now with these definitions, any vector whatsoever can be written this way:

$$\mathbf{a} = a_x \mathbf{i} + a_y \mathbf{j} + a_z \mathbf{k}. \tag{1.25}$$

By this means we can go from the components of a vector to the vector itself.

This discussion of vectors is by no means complete. However, rather than try to go more deeply into the subject now, we shall first learn to use in physical situations some of the ideas so far discussed. Then, when we have properly mastered this basic material, we shall find it easier to penetrate more deeply into the subject without getting too confused. We shall later find that it is useful to define another kind of product of two vectors, called the vector product, and written as $\mathbf{a} \times \mathbf{b}$. However, we shall undertake a discussion of such matters in a later chapter.

Two

SYMMETRY IN PHYSICAL LAWS

2-1 Symmetry operations

◊ The subject of this chapter is what we may call *symmetry in physical laws*. We have already discussed certain features of symmetry in physical laws in connection with vector analysis (Chapter 1), the theory of relativity (which follows in Chapter 4), and rotation (Chapter 20*).

Why should we be concerned with symmetry? In the first place, symmetry is fascinating to the human mind, and everyone likes objects or patterns that are in some way symmetrical. It is an interesting fact that nature often exhibits certain kinds of symmetry in the objects we find in the world around us. Perhaps the most symmetrical object imaginable is a sphere, and nature is full of spheres—stars, planets, water droplets in clouds. The crystals found in rocks exhibit many different kinds of symmetry, the study of which tells us some important things about the structure of solids. Even the animal and vegetable worlds show some degree of symmetry, although the symmetry of a flower or of a bee is not as perfect or as fundamental as is that of a crystal.

But our main concern here is not with the fact that the *objects* of

---

* of the original *Lectures on Physics*, vol. I.

nature are often symmetrical. Rather, we wish to examine some of the even more remarkable symmetries of the universe—the symmetries that exist in the *basic laws themselves* which govern the operation of the physical world.

First, what *is* symmetry? How can a physical *law* be "symmetrical"? The problem of defining symmetry is an interesting one and we have already noted that Weyl gave a good definition, the substance of which is that a thing is symmetrical if there is something we can do to it so that after we have done it, it looks the same as it did before. For example, a symmetrical vase is of such a kind that if we reflect or turn it, it will look the same as it did before. The question we wish to consider here is what we can do to physical phenomena, or to a physical situation in an experiment, and yet leave the result the same. A list of the known operations under which various physical phenomena remain invariant is shown in Table 2-1.

2-2 *Symmetry in space and time*

The first thing we might try to do, for example, is to *translate* the phenomenon in space. If we do an experiment in a certain region, and then build another apparatus at another place in space (or move the original one over) then, whatever went on in one apparatus, in a certain order in time, will occur in the same way if we

Table 2-1. Symmetry Operations

Translation in space
Translation in time
Rotation through a fixed angle
Uniform velocity in a straight line (Lorentz transformation)
Reversal of time
Reflection of space
Interchange of identical atoms or identical particles
Quantum-mechanical phase
Matter-antimatter (charge conjugation)

have arranged the same condition, with all due attention to the restrictions that we mentioned before: that all of those features of the environment which make it not behave the same way have also been moved over—we talked about how to define how much we should include in those circumstances, and we shall not go into those details again.

In the same way, we also believe today that *displacement in time* will have no effect on physical laws. (That is, *as far as we know today*—all of these things are as far as we know today!) That means that if we build a certain apparatus and start it at a certain time, say on Thursday at 10:00 A.M., and then build the same apparatus and start it, say, three days later in the same condition, the two apparatuses will go through the same motions in exactly the same way as a function of time no matter what the starting time, provided again, of course, that the relevant features of the environment are also modified appropriately in *time*. That symmetry means, of course, that if one bought General Motors stock three months ago, the same thing would happen to it if he bought it now!

We have to watch out for geographical differences too, for there are, of course, variations in the characteristics of the earth's surface. So, for example, if we measure the magnetic field in a certain region and move the apparatus to some other region, it may not work in precisely the same way because the magnetic field is different, but we say that is because the magnetic field is associated with the earth. We can imagine that if we move the whole earth and the equipment, it would make no difference in the operation of the apparatus.

Another thing that we discussed in considerable detail was rotation in space: if we turn an apparatus at an angle it works just as well, provided we turn everything else that is relevant along with it. In fact, we discussed the problem of symmetry under rotation in space in some detail in Chapter 1, and we invented a mathematical system called *vector analysis* to handle it as neatly as possible.

On a more advanced level we had another symmetry—the symmetry under uniform velocity in a straight line. That is to say—a rather remarkable effect—that if we have a piece of apparatus working a certain way and then take the same apparatus and put it in a car, and move the whole car, plus all the relevant surroundings, at a uniform velocity in a straight line, then so far as the phenomena inside the car are concerned there is no difference: all the laws of physics appear the same. We even know how to express this more technically, and that is that the mathematical equations of the physical laws must be unchanged under a *Lorentz transformation*. As a matter of fact, it was a study of the relativity problem that concentrated physicists' attention most sharply on symmetry in physical laws.

Now the above-mentioned symmetries have all been of a geometrical nature, time and space being more or less the same, but there are other symmetries of a different kind. For example, there is a symmetry which describes the fact that we can replace one atom by another of the same kind; to put it differently, there *are* atoms of the same kind. It is possible to find groups of atoms such that if we change a pair around, it makes no difference—the atoms are identical. Whatever one atom of oxygen of a certain type will do, another atom of oxygen of that type will do. One may say, "That is ridiculous, that is the *definition* of equal types!" That may be merely the definition, but then we still do not know whether there *are* any "atoms of the same type"; the *fact* is that there are many, many atoms of the same type. Thus it does mean something to say that it makes no difference if we replace one atom by another of the same type. The so-called elementary particles of which the atoms are made are also identical particles in the above sense—all electrons are the same; all protons are the same; all positive pions are the same; and so on.

After such a long list of things that can be done without changing the phenomena, one might think we could do practically anything; so let us give some examples to the contrary, just to see the difference. Suppose that we ask: "Are the physical laws symmetrical under a change of scale?" Suppose we build a certain piece of apparatus, and then build another apparatus five times bigger in every part, will it work exactly the same way? The answer is, in this

Symmetry in Physical Laws

case, *no*! The wavelength of light emitted, for example, by the atoms inside one box of sodium atoms and the wavelength of light emitted by a gas of sodium atoms five times in volume is not five times longer, but is in fact exactly the same as the other. So the ratio of the wavelength to the size of the emitter will change.

Another example: we see in the newspaper, every once in a while, pictures of a great cathedral made with little matchsticks—a tremendous work of art by some retired fellow who keeps gluing matchsticks together. It is much more elaborate and wonderful than any real cathedral. If we imagine that this wooden cathedral were actually built on the scale of a real cathedral, we see where the trouble is; it would not last—the whole thing would collapse because of the fact that scaled-up matchsticks are just not strong enough. "Yes," one might say, "but we also know that when there is an influence from the outside, it also must be changed in proportion!" We are talking about the ability of the object to withstand gravitation. So what we should do is first to take the model cathedral of real matchsticks and the real earth, and then we know it is stable. Then we should take the larger cathedral and take a bigger earth. But then it is even worse, because the gravitation is increased still more!

Today, of course, we understand the fact that phenomena depend on the scale on the grounds that matter is atomic in nature, and certainly if we built an apparatus that was so small there were only five atoms in it, it would clearly be something we could not scale up and down arbitrarily. The scale of an individual atom is not at all arbitrary—it is quite definite.

The fact that the laws of physics are not unchanged under a change of scale was discovered by Galileo. He realized that the strengths of materials were not in exactly the right proportion to their sizes, and he illustrated this property that we were just discussing, about the cathedral of matchsticks, by drawing two bones, the bone of one dog, in the right proportion for holding up his weight, and the imaginary bone of a "super dog" that would be, say, ten or a hundred times bigger—that bone was a big, solid thing with quite different proportions. We do not know whether he ever carried the argument quite to the conclusion that the laws of nature must have a definite scale, but he was so impressed with this discovery that he considered it to be as important as the discovery of the laws of motion, because he published them both in the same volume, called "On Two New Sciences."

Another example in which the laws are not symmetrical, that we know quite well, is this: a system in rotation at a uniform angular velocity does not give the same apparent laws as one that is not rotating. If we make an experiment and then put everything in a space ship and have the space ship spinning in empty space, all alone at a constant angular velocity, the apparatus will not work the same way because, as we know, things inside the equipment will be thrown to the outside, and so on, by the centrifugal or coriolis forces, etc. In fact, we can tell that the earth is rotating by using a so-called Foucault pendulum, without looking outside.

Next we mention a very interesting symmetry which is obviously false, i.e., *reversibility in time*. The physical laws apparently cannot be reversible in time, because, as we know, all obvious phenomena are irreversible on a large scale: "The moving finger writes, and having writ, moves on." So far as we can tell, this irreversibility is due to the very large number of particles involved, and if we could see the individual molecules, we would not be able to discern whether the machinery was working forwards or backwards. To make it more precise: we build a small apparatus in which we know what all the atoms are doing, in which we can watch them jiggling. Now we build another apparatus like it, but which starts its motion in the final condition of the other one, with all the velocities precisely reversed. *It will then go through the same motions, but exactly in reverse.* Putting it another way: if we take a motion picture, with sufficient detail, of all the inner works of a piece of material and shine it on a screen and run it backwards, no physicist will be able to say, "That is against the laws of physics, that is doing something wrong!" If we do not see all the details, of course, the situation will be perfectly clear. If we see the egg splattering on the sidewalk and the shell cracking open, and so on, then we will surely say, "That is irreversible, because if we run the moving picture backwards the egg will all collect together and the shell will go back together, and that is obviously ridiculous!" But if we look at the individual atoms themselves, the laws look completely reversible. This is, of course, a much harder discovery to have made, but apparently it is true that the fundamental physical laws, on a microscopic and fundamental level, are completely reversible in time!

2-3 Symmetry and conservation laws

The symmetries of the physical laws are very interesting at this level, but they turn out, in the end, to be even more interesting and exciting when we come to quantum mechanics. For a reason which we cannot make clear at the level of the present discussion—a fact that most physicists still find somewhat staggering, a most profound and beautiful thing, is that, in quantum mechanics, *for each of the rules of symmetry there is a corresponding conservation law*; there is a definite connection between the laws of conservation and the symmetries of physical laws. We can only state this at present, without any attempt at explanation.

The fact, for example, that the laws are symmetrical for translation in space when we add the principles of quantum mechanics, turns out to mean that *momentum is conserved.*

That the laws are symmetrical under translation in time means, in quantum mechanics, that *energy is conserved.*

Invariance under rotation through a fixed angle in space corresponds to the *conservation of angular momentum.* These connections are very interesting and beautiful things, among the most beautiful and profound things in physics.

Incidentally, there are a number of symmetries which appear in quantum mechanics which have no classical analog, which have no method of description in classical physics. One of these is as follows: If ψ is the amplitude for some process or other, we know that the absolute square of ψ is the probability that the process will occur. Now if someone else were to make his calculations, not with this ψ, but with a ψ' which differs merely by a change in phase (let Δ be some constant, and multiply $e^{i\Delta}$ times the old ψ), the absolute square of ψ', which is the probability of the event, is then equal to the absolute square of ψ:

$$\psi' = \psi e^{i\Delta}; \quad |\psi'|^2 = |\psi|^2. \tag{2.1}$$

Therefore the physical laws are unchanged if the phase of the wave function is shifted by an arbitrary constant. That is another symmetry. Physical laws must be of such a nature that a shift in the quantum-mechanical phase makes no difference. As we have just mentioned, in quantum mechanics there is a conservation law for every symmetry. The conservation law which is connected with the quantum-mechanical phase seems to be the *conservation of electrical charge*. This is altogether a very interesting business!

2-4 Mirror reflections

Now the next question, which is going to concern us for most of the rest of this chapter, is the question of symmetry under *reflection in space*. The problem is this: Are the physical laws symmetrical under reflection? We may put it this way: Suppose we build a piece of equipment, let us say a clock, with lots of wheels and hands and numbers; it ticks, it works, and it has things wound up inside. We look at the clock in the mirror. How it *looks* in the mirror is not the question. But let us actually *build* another clock which is exactly the same as the way the first clock looks in the mirror—every time there is a screw with a right-hand thread in one, we use a screw with a left-hand thread in the corresponding place of the other; where one is marked "2" on the face, we mark a "Σ" on the face of the other; each coiled spring is twisted one way in one clock and the other way in the mirror-image clock; when we are all finished, we have two clocks, both physical, which bear to each other the relation of an object and its mirror image, although they are both actual, material objects, we emphasize. Now the question is: If the two clocks are

Symmetry in Physical Laws

started in the same condition, the springs wound to corresponding tightnesses, will the two clocks tick and go around, forever after, as exact mirror images? (This is a physical question, not a philosophical question.) Our intuition about the laws of physics would suggest that they *would*.

We would suspect that, at least in the case of these clocks, reflection in space is one of the symmetries of physical laws, that if we change everything from "right" to "left" and leave it otherwise the same, we cannot tell the difference. Let us, then, suppose for a moment that this is true. If it is true, then it would be impossible to distinguish "right" and "left" by any physical phenomenon, just as it is, for example, impossible to define a particular absolute velocity by a physical phenomenon. So it should be impossible, by any physical phenomenon, to define absolutely what we mean by "right" as opposed to "left," because the physical laws should be symmetrical.

Of course, the world does not *have* to be symmetrical. For example, using what we may call "geography," surely "right" can be defined. For instance, we stand in New Orleans and look at Chicago, and Florida is to our right (when our feet are on the ground!). So we can define "right" and "left" by geography. Of course, the actual situation in any system does not have to have the symmetry that we are talking about; it is a question of whether the *laws* are symmetrical—in other words, whether it is *against the physical laws* to have a sphere like the earth with "left-handed dirt" on it and a person like ourselves standing looking at a city like Chicago from a place like New Orleans, but with everything the other way around, so Florida is on the other side. It clearly seems not impossible, not against the physical laws, to have everything changed left for right.

Another point is that our definition of "right" should not depend on history. An easy way to distinguish right from left is to go to a machine shop and pick up a screw at random. The odds are it has a right-hand thread—not necessarily, but it is much more likely to have a right-hand thread than a left-hand one. This is a question of history or convention, or the way things happen to be, and is again not a question of fundamental laws. As we can well appreciate, everyone could have started out making left-handed screws!

So we must try to find some phenomenon in which "right hand" is involved fundamentally. The next possibility we discuss is the fact that polarized light rotates its plane of polarization as it goes through, say, sugar water. As we saw in Chapter 33,* it rotates, let us say, to the right in a certain sugar solution. That is a way of defining "right-hand," because we may dissolve some sugar in the water and then the polarization goes to the right. But sugar has come from living things, and if we try to make the sugar artificially, then we discover that it *does not* rotate the plane of polarization! But if we then take that same sugar which is made artificially and which does not rotate the plane of polarization, and put bacteria in it (they eat some of the sugar) and then filter out the bacteria, we find that we still have sugar left (almost half as much as we had before), and this time it does rotate the plane of polarization, but *the other way*! It seems very confusing, but is easily explained.

Take another example: One of the substances which is common to all living creatures and that is fundamental to life is protein. Proteins consist of chains of amino acids. Figure 2-1 shows a model of an amino acid that comes out of a protein. This amino acid is called alanine, and the molecular arrangement would look like that in Figure 2-1(a) if it came out of a protein of a real living thing. On the other hand, if we try to make alanine from carbon dioxide, ethane, and ammonia (and we *can* make it, it is not a complicated molecule), we discover that we are making equal amounts of this molecule and the one shown in Figure 2-1(b)! The first molecule, the one that comes from the living thing, is called *L-alanine*. The other one, which is the same chemically, in that it has the same kinds of atoms and the same connections of the atoms, is a "right-hand" molecule, compared with the "left-hand" L-alanine, and it is called *D-alanine*. The interesting thing is that when we make alanine at

* of the original *Lectures on Physics*, vol. I.

Symmetry in Physical Laws

Figure 2-1. (a) L-alanine (left), and (b) D-alanine (right).

home in a laboratory from simple gases, we get an equal mixture of both kinds. However, the only thing that life uses is L-alanine. (This is not exactly true. Here and there in living creatures there is a special use for D-alanine, but it is very rare. All proteins use L-alanine exclusively.) Now if we make both kinds, and we feed the mixture to some animal which likes to "eat," or use up, alanine, it cannot use D-alanine, so it only uses the L-alanine; that is what happened to our sugar—after the bacteria eat the sugar that works well for them, only the "wrong" kind is left! (Left-handed sugar tastes sweet, but not the same as right-handed sugar.)

So it looks as though the phenomena of life permit a distinction between "right" and "left," or chemistry permits a distinction, because the two molecules are chemically different. But no, it does not! So far as physical measurements can be made, such as of energy, the rates of chemical reactions, and so on, the two kinds work exactly the same way if we make everything else in a mirror image too. One molecule will rotate light to the right, and the other will rotate it to the left in precisely the same amount, through the same amount of fluid. Thus, so far as physics is concerned, these two amino acids are equally satisfactory. So far as we understand things today, the fundamentals of the Schrödinger equation have it that the two molecules should behave in exactly corresponding ways, so that one is to the right as the other is to the left. Neverthe-less, in life it is all one way!

It is presumed that the reason for this is the following. Let us suppose, for example, that life is somehow at one moment in a certain condition in which all the proteins in some creatures have left-handed amino acids, and all the enzymes are lopsided—every substance in the living creature is lopsided—it is not symmetrical. So when the digestive enzymes try to change the chemicals in the food from one kind to another, one kind of chemical "fits" into the enzyme, but the other kind does not (like Cinderella and the slipper, except that it is a "left foot" that we are testing). So far as we know, in principle, we could build a frog, for example, in which every molecule is reversed, everything is like the "left-hand" mirror image of a real frog; we have a left-hand frog. This left-hand frog would go on all right for a while, but he would find nothing to eat, because if he swallows a fly, his enzymes are not built to digest it. The fly has the wrong "kind" of amino acids (unless we give him a left-hand fly). So as far as we know, the chemical and life processes would con-tinue in the same manner if everything were reversed.

If life is entirely a physical and chemical phenomenon, then we can understand that the proteins are all made in the same corkscrew only from the idea that at the very beginning some living molecules, by accident, got started and a few won. Somewhere, once, one organic molecule was lopsided in a certain way, and from this particular thing the "right" happened to evolve in our particular geography; a particular historical accident was one-sided, and ever since then the lopsidedness has propagated itself. Once having arrived at the state that it is in now, of course, it will always continue—all the enzymes digest the right things, manufacture the right things: when the carbon dioxide and the water vapor, and so on, go in the plant leaves, the enzymes that make the sugars make them lopsided because the enzymes are lopsided. If any new kind of virus or living thing were to originate at a later time, it would survive only if it could "eat" the kind of living matter already present. Thus it, too, must be of the same kind.

Symmetry in Physical Laws

There is no conservation of the number of right-handed molecules. Once started, we could keep increasing the number of right-handed molecules. So the presumption is, then, that the phenomena in the case of life do not show a lack of symmetry in physical laws, but do show, on the contrary, the universal nature and the commonness of ultimate origin of all creatures on earth, in the sense described above.

2-5 Polar and axial vectors

Now we go further. We observe that in physics there are a lot of other places where we have "right-hand" and "left-hand" rules. As a matter of fact, when we learned about vector analysis we learned about the right-hand rules we have to use in order to get the angular momentum, torque, magnetic field, and so on, to come out right. The force on a charge moving in a magnetic field, for example, is $\mathbf{F} = q\mathbf{v} \times \mathbf{B}$. In a given situation, in which we know $\mathbf{F}$, $\mathbf{v}$, and $\mathbf{B}$, isn't that equation enough to define right-handedness? As a matter of fact, if we go back and look at where the vectors came from, we know that the "right-hand rule" was merely a convention; it was a trick. The original quantities, like the angular momenta and the angular velocities, and things of this kind, were not really vectors at all! They are all somehow associated with a certain plane, and it is just because there are three dimensions in space that we can associate the quantity with a direction perpendicular to that plane. Of the two possible directions, we chose the "right-hand" direction.

So if the laws of physics are symmetrical, we should find that if some demon were to sneak into all the physics laboratories and replace the word "right" for "left" in every book in which "right-hand rules" are given, and instead we were to use all "left-hand rules," uniformly, then it should make no difference whatever in the physical laws.

Let us give an illustration. There are two kinds of vectors. There are "honest" vectors, for example a step $\Delta\mathbf{r}$ in space. If in our apparatus there is a piece here and something else there, then in a mirror apparatus there will be the image piece and the image something else, and if we draw a vector from the "piece" to the "something else," one vector is the mirror image of the other (Figure 2-2). The vector arrow changes its head, just as the whole space turns inside out; such a vector we call a *polar vector*.

But the other kind of vector, which has to do with rotations, is of a different nature. For example, suppose that in three dimensions something is rotating as shown in Figure 2-3. Then if we look at it in a mirror, it will be rotating as indicated, namely, as the mirror image of the original rotation. Now we have agreed to represent the mirror rotation by the same rule, it is a "vector" which, on reflection, does *not* change about as the polar vector does, but is reversed relative to the polar vectors and to the geometry of the space; such a vector is called an *axial vector*.

Now if the law of reflection symmetry is right in physics, then it must be true that the equations must be so designed that if we change the sign of each axial vector and each cross-product of vectors, which would be what corresponds to reflection, nothing will happen. For instance, when we write a formula which says that the angular momentum is $\mathbf{L} = \mathbf{r} \times \mathbf{p}$, that equation is all right, because if we change to a left-hand coordinate system, we change the sign of $\mathbf{L}$, but $\mathbf{p}$ and $\mathbf{r}$ do not change; the cross-product sign is changed, since we must change from a right-hand rule to a left-hand rule. As another example, we know that the force on a charge moving in a magnetic field is $\mathbf{F} = q\mathbf{v} \times \mathbf{B}$, but if we change from a right- to a left-handed system, since $\mathbf{F}$ and $\mathbf{v}$ are known to be polar vectors the sign change required by the cross-product must be cancelled by a sign change in $\mathbf{B}$, which means that $\mathbf{B}$ must be an

Figure 2-2. A step in space and its mirror image.

Symmetry in Physical Laws

Figure 2-3. A rotating wheel and its mirror image. Note that the angular velocity "vector" is not reversed in direction.

axial vector. In other words, if we make such a reflection, **B** must go to − **B**. So if we change our coordinates from right to left, we must also change the poles of magnets from north to south.

Let us see how that works in an example. Suppose that we have two magnets, as in Figure 2-4. One is a magnet with the coils going around a certain way, and with current in a given direction. The other magnet looks like the reflection of the first magnet in a mirror—the coil will wind the other way, everything that happens inside the coil is exactly reversed, and the current goes as shown. Now, from the laws for the production of magnetic fields, which we do not know yet officially, but which we most likely learned in high school, it turns out that the magnetic field is as shown in the figure. In one case the pole is a south magnetic pole, while in the other magnet the current is going the other way and the magnetic field is reversed—it is a north magnetic pole. So we see that when we go from right to left we must indeed change from north to south!

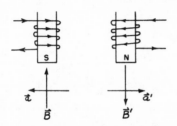

Figure 2-4. A magnet and its mirror image.

Never mind changing north to south; these too are mere conventions. Let us talk about *phenomena*. Suppose, now, that we have an electron moving through one field, going into the page. Then, if we use the formula for the force, $\mathbf{v} \times \mathbf{B}$ (remember the charge is minus), we find that the electron will deviate in the indicated direction according to the physical law. So the phenomenon is that we have a coil with a current going in a specified sense and an electron curves in a certain way—that is the physics—never mind how we label everything.

Now let us do the same experiment with a mirror: we send an electron through in a corresponding direction and now the force is reversed, if we calculate it from the same rule, and that is very good because the corresponding *motions* are then mirror images!

2-6 *Which hand is right?*

So the fact of the matter is that in studying any phenomenon there are always two right-hand rules, or an even number of them, and the net result is that the phenomena always look symmetrical. In short, therefore, we cannot tell right from left if we also are not able to tell north from south. However, it may seem that we *can* tell the north pole of a magnet. The north pole of a compass needle, for example, is one that points to the north. But of course that is again a local property that has to do with geography of the earth; that is just like talking about in which direction is Chicago, so it does not count. If we have seen compass needles, we may have noticed that the north-seeking pole is a sort of bluish color. But that is just due to the man who painted the magnet. These are all local, conventional criteria.

However, if a magnet were to have the property that if we looked at it closely enough we would see small hairs growing on its north pole but not on its south pole, if that were the general rule, or if there were *any* unique way to distinguish the north from the south pole of a magnet, then we could tell which of the two cases we actually had, and *that would be the end of the law of reflection symmetry.*

Symmetry in Physical Laws

To illustrate the whole problem still more clearly, imagine that we were talking to a Martian, or someone very far away, by telephone. We are not allowed to send him any actual samples to inspect; for instance, if we could send light, we could send him right-hand circularly polarized light and say, "That is right-hand light—just watch the way it is going." But we cannot *give* him anything, we can only talk to him. He is far away, or in some strange location, and he cannot see anything we can see. For instance, we cannot say, "Look at Ursa major; now see how those stars are arranged. What we mean by 'right' is . . ." We are only allowed to telephone him.

Now we want to tell him all about us. Of course, first we start defining numbers, and say, "Tick, tick, *two*, tick, tick, tick, *three* . . . ," so that gradually he can understand a couple of words, and so on. After a while we may become very familiar with this fellow, and he says, "What do you guys look like?" We start to describe ourselves, and say, "Well, we are six feet tall." He says, "Wait a minute, what is six feet?" Is it possible to tell him what six feet is? Certainly! We say, "You know about the diameter of hydrogen atoms—we are 17,000,000,000 hydrogen atoms high!" That is possible because physical laws are not variant under change of scale, and therefore we *can* define an absolute length. And so we define the size of the body, and tell him what the general shape is—it has prongs with five bumps sticking out on the ends, and so on, and he follows us along, and we finish describing how we look on the outside, presumably without encountering any particular difficulties. He is even making a model of us as we go along. He says, "My, you are certainly very handsome fellows; now what is on the inside?" So we start to describe the various organs on the inside, and we come to the heart, and we carefully describe the shape of it, and say, "Now put the heart on the left side." He says, "Duhhh—the left side?" Now our problem is to describe to him which side the heart goes on without his ever seeing anything that we see, and without our ever sending any sample to him of what we mean by "right"—no standard right-handed object. Can we do it?

2-7 Parity is not conserved!

It turns out that the laws of gravitation, the laws of electricity and magnetism, nuclear forces, all satisfy the principle of reflection symmetry, so these laws, or anything derived from them, cannot be used. But associated with the many particles that are found in nature there is a phenomenon called *beta decay*, or *weak decay*. One of the examples of weak decay, in connection with a particle discovered in about 1954, posed a strange puzzle. There was a certain charged particle which disintegrated into three π-mesons, as shown schematically in Figure 2-5. This particle was called, for a while, a τ-meson. Now in Figure 2-5 we also see another particle which disintegrates into *two* mesons; one must be neutral, from the conservation of charge. This particle was called a θ-meson. So on the one hand we have a particle called a τ, which disintegrates into three π-mesons, and a θ, which disintegrates into two π-mesons. Now it was soon discovered that the τ and the θ are almost equal in mass; in fact, within the experimental error, they are equal. Next, the length of time it took for them to disintegrate into three π's and two π's was found to be almost exactly the same; they live the same length of time. Next, whenever they were made, they were made in the same proportions, say, 14 percent τ's to 86 percent θ's.

Anyone in his right mind realizes immediately that they must be the same particle, that we merely produce an object which has two different ways of disintegrating—not two different particles. This object that can disintegrate in two different ways has, therefore,

Figure 2-5. A schematic diagram of the disintegration of a τ^+ and a θ^+ particle.

Symmetry in Physical Laws

the same lifetime and the same production ratio (because this is simply the ratio of the odds with which it disintegrates into these two kinds).

However, it was possible to prove (and we cannot here explain at all *how*), from the principle of reflection symmetry in quantum mechanics, that it was *impossible* to have these both come from the same particle—the same particle *could not* disintegrate in both of these ways. The conservation law corresponding to the principle of reflection symmetry is something which has no classical analog, and so this kind of quantum-mechanical conservation was called the *conservation of parity*. So, it was a result of the conservation of parity or, more precisely, from the symmetry of the quantum-mechanical equations of the weak decays under reflection, that the same particle could not go into both, so it must be some kind of coincidence of masses, lifetimes, and so on. But the more it was studied, the more remarkable the coincidence, and the suspicion gradually grew that possibly the deep law of the reflection symmetry of nature may be false.

As a result of this apparent failure, the physicists Lee and Yang suggested that other experiments be done in related decays to try to test whether the law was correct in other cases. The first such experiment was carried out by Miss Wu from Columbia, and was done as follows. Using a very strong magnet at a very low temperature, it turns out that a certain isotope of cobalt, which disintegrates by emitting an electron, is magnetic, and if the temperature is low enough that the thermal oscillations do not jiggle the atomic magnets about too much, they line up in the magnetic field. So the cobalt atoms will all line up in this strong field. They then disintegrate, emitting an electron, and it was discovered that when the atoms were lined up in a field whose **B** vector points upward, most of the electrons were emitted in a downward direction.

If one is not really "hep" to the world, such a remark does not sound like anything of significance, but if one appreciates the problems and interesting things in the world, then he sees that it is a most dramatic discovery: When we put cobalt atoms in an extremely strong magnetic field, more disintegration electrons go down than up. Therefore if we were to put it in a corresponding experiment in a "mirror," in which the cobalt atoms would be lined up in the opposite direction, they would spit their electrons *up*, not *down*; the action is *unsymmetrical. The magnet has grown hairs!* The south pole of a magnet is of such a kind that the electrons in a β-disintegration tend to go away from it; that distinguishes, in a physical way, the north pole from the south pole.

After this, a lot of other experiments were done: the disintegration of the π into μ and ν; μ into an electron and two neutrinos; nowadays, the Λ into proton and π; disintegration of Σ's; and many other disintegrations. In fact, in almost all cases where it could be expected, all have been found *not* to obey reflection symmetry! Fundamentally, the law of reflection symmetry, at this level in physics, is incorrect.

In short, we can tell a Martian where to put the heart: we say, "Listen, build yourself a magnet, and put the coils in, and put the current on, and then take some cobalt and lower the temperature. Arrange the experiment so the electrons go from the foot to the head, then the direction in which the current goes through the coils is the direction that goes in on what we call the right and comes out on the left." So it is possible to define right and left, now, by doing an experiment of this kind.

There are a lot of other features that were predicted. For example, it turns out that the spin, the angular momentum, of the cobalt nucleus before disintegration is 5 units of $\hbar$, and after disintegration it is 4 units. The electron carries spin angular momentum, and there is also a neutrino involved. It is easy to see from this that the electron must carry its spin angular momentum aligned along its direction of motion, the neutrino likewise. So it looks as though the electron is spinning to the left, and that was also checked. In fact, it was checked right here at Caltech by Boehm and Wapstra, that the electrons spin mostly to the left.

(There were some other experiments that gave the opposite answer, but they were wrong!)

The next problem, of course, was to find the law of the failure of parity conservation. What is the rule that tells us how strong the failure is going to be? The rule is this: it occurs only in these very slow reactions, called weak decays, and when it occurs, the rule is that the particles which carry spin, like the electron, neutrino, and so on, come out with a spin tending to the left. That is a lopsided rule; it connects a polar vector velocity and an axial vector angular momentum, and says that the angular momentum is more likely to be opposite to the velocity than along it.

Now that is the rule, but today we do not really understand the whys and wherefores of it. *Why* is this the right rule, what is the fundamental reason for it, and how is it connected to anything else? At the moment we have been so shocked by the fact that this thing is unsymmetrical that we have not been able to recover enough to understand what it means with regard to all the other rules. However, the subject is interesting, modern, and still unsolved, so it seems appropriate that we discuss some of the questions associated with it.

2-8 Antimatter

The first thing to do when one of the symmetries is lost is to immediately go back over the list of known or assumed symmetries and ask whether any of the others are lost. Now we did not mention one operation on our list, which must necessarily be questioned, and that is the relation between matter and antimatter. Dirac predicted that in addition to electrons there must be another particle, called the positron (discovered at Caltech by Anderson), that is necessarily related to the electron. All the properties of these two particles obey certain rules of correspondence: the energies are equal; the masses are equal; the charges are reversed; but, more important than anything, the two of them, when they come together, can annihilate each other and liberate their entire mass in the form of energy, say γ-rays. The positron is called an *antiparticle* to the electron, and these are the characteristics of a particle and its antiparticle. It was clear from Dirac's argument that all the rest of the particles in the world should also have corresponding antiparticles. For instance, for the proton there should be an antiproton, which is now symbolized by a $\bar{p}$. The $\bar{p}$ would have a negative electrical charge and the same mass as a proton, and so on. The most important feature, however, is that a proton and an antiproton coming together can annihilate each other. The reason we emphasize this is that people do not understand it when we say there is a neutron and also an antineutron, because they say, "A neutron is neutral, so how *can* it have the opposite charge?" The rule of the "anti" is not just that it has the opposite charge, it has a certain set of properties, the whole lot of which are opposite. The antineutron is distinguished from the neutron in this way: if we bring two neutrons together, they just stay as two neutrons, but if we bring a neutron and an antineutron together, they annihilate each other with a great explosion of energy being liberated, with various π-mesons, γ-rays, and whatnot.

Now if we have antineutrons, antiprotons, and antielectrons, we can make antiatoms, in principle. They have not been made yet, but it is possible in principle. For instance, a hydrogen atom has a proton in the center with an electron going around outside. Now imagine that somewhere we can make an antiproton with a positron going around, would it go around? Well, first of all, the antiproton is electrically negative and the antielectron is electrically positive, so they attract each other in a corresponding manner—the masses are all the same; everything is the same. It is one of the principles of the symmetry of physics, the equations seem to show, that if a clock, say, were made of matter on one hand, and then we made the same clock of antimatter, it would run in this way. (Of course, if we put the clocks together, they would annihilate each other, but that is different.)

An immediate question then arises. We can build, out of matter,

Symmetry in Physical Laws

two clocks, one which is "left-hand" and one which is "right-hand." For example, we could build a clock which is not built in a simple way, but has cobalt and magnets and electron detectors which detect the presence of β-decay electrons and count them. Each time one is counted, the second hand moves over. Then the mirror clock, receiving fewer electrons, will not run at the same rate. So evidently we can make two clocks such that the left-hand clock does not agree with the right-hand one. Let us make, out of matter, a clock which we call the standard or right-hand clock. Now let us make, also out of matter, a clock which we call the left-hand clock. We have just discovered that, in general, these two will *not* run the same way; prior to that famous physical discovery, it was thought that they would. Now it was also supposed that matter and antimatter were equivalent. That is, if we made an antimatter clock, right-hand, the same shape, then it would run the same as the right-hand matter clock, and if we made the same clock to the left it would run the same. In other words, in the beginning it was believed that *all four* of these clocks were the same; now of course we know that the right-hand and left-hand matter are not the same. Presumably, therefore, the right-handed antimatter and the left-handed antimatter are not the same.

So the obvious question is, which goes with which, if either? In other words, does the right-handed matter behave the same way as the right-handed antimatter? Or does the right-handed matter behave the same way as the left-handed antimatter? β-decay experiments, using positron decay instead of electron decay, indicate that this is the interconnection: matter to the "right" works the same way as antimatter to the "left."

Therefore, at long last, it is really true that right and left symmetry is still maintained! If we made a left-hand clock, but made it out of the other kind of matter, antimatter instead of matter, it would run in the same way. So what has happened is that instead of having two independent rules in our list of symmetries, two of these rules go together to make a new rule, which says that matter to the right is symmetrical with antimatter to the left.

So if our Martian is made of antimatter and we give him instructions to make this "right" handed model like us, it will, of course, come out the other way around. What would happen when, after much conversation back and forth, we each have taught the other to make space ships and we meet halfway in empty space? We have instructed each other on our traditions, and so forth, and the two of us come rushing out to shake hands. Well, if he puts out his left hand, watch out!

2-9 Broken symmetries

The next question is, what can we make out of laws which are *nearly* symmetrical? The marvelous thing about it all is that for such a wide range of important, strong phenomena—nuclear forces, electrical phenomena, and even weak ones like gravitation—over a tremendous range of physics, all the laws for these seem to be symmetrical. On the other hand, this little extra piece says, "No, the laws are not symmetrical!" How is it that nature can be almost symmetrical, but not perfectly symmetrical? What shall we make of this? First, do we have any other examples? The answer is, we do, in fact, have a few other examples. For instance, the nuclear part of the force between proton and proton, between neutron and neutron, and between neutron and proton, is all exactly the same—there is a symmetry for nuclear forces, a new one, that we can interchange neutron and proton—but it evidently is not a general symmetry, for the electrical repulsion between two protons at a distance does not exist for neutrons. So it is not generally true that we can *always* replace a proton with a neutron, but only to a good approximation. Why *good*? Because the nuclear forces are much stronger than the electrical forces. So this is an "almost" symmetry also. So we do have examples in other things.

We have, in our minds, a tendency to accept symmetry as some kind of perfection. In fact it is like the old idea of the Greeks that circles were perfect, and it was rather horrible to believe that the planetary orbits were not circles, but only nearly circles. The differ-

Symmetry in Physical Laws

ence between being a circle and being nearly a circle is not a small difference, it is a fundamental change so far as the mind is concerned. There is a sign of perfection and symmetry in a circle that is not there the moment the circle is slightly off—that is the end of it—it is no longer symmetrical. Then the question is why it is only *nearly* a circle—that is a much more difficult question. The actual motion of the planets, in general, should be ellipses, but during the ages, because of tidal forces, and so on, they have been made almost symmetrical. Now the question is whether we have a similar problem here. The problem from the point of view of the circles is if they were perfect circles there would be nothing to explain, that is clearly simple. But since they are only nearly circles, there is a lot to explain, and the result turned out to be a big dynamical problem, and now our problem is to explain why they are nearly symmetrical by looking at tidal forces and so on.

So our problem is to explain where symmetry comes from. Why is nature so nearly symmetrical? No one has any idea why. The only thing we might suggest is something like this: There is a gate in Japan, a gate in Neiko, which is sometimes called by the Japanese the most beautiful gate in all Japan; it was built in a time when there was great influence from Chinese art. This gate is very elaborate, with lots of gables and beautiful carving and lots of columns and dragon heads and princes carved into the pillars, and so on. But when one looks closely he sees that in the elaborate and complex design along one of the pillars, one of the small design elements is carved upside down; otherwise the thing is completely symmetrical. If one asks why this is, the story is that it was carved upside down so that the gods will not be jealous of the perfection of man. So they purposely put an error in there, so that the gods would not be jealous and get angry with human beings.

We might like to turn the idea around and think that the true explanation of the near symmetry of nature is this: that God made the laws only nearly symmetrical so that we should not be jealous of His perfection!

Three

THE SPECIAL THEORY
OF RELATIVITY

3-1 The principle of relativity

For over 200 years the equations of motion enunciated by Newton were believed to describe nature correctly, and the first time that an error in these laws was discovered, the way to correct it was also discovered. Both the error and its correction were discovered by Einstein in 1905.

Newton's Second Law, which we have expressed by the equation

$$F = d(mv)/dt,$$

was stated with the tacit assumption that m is a constant, but we now know that this is not true, and that the mass of a body increases with velocity. In Einstein's corrected formula m has the value

$$m = \frac{m_0}{\sqrt{1 - v^2/c^2}}, \tag{3.1}$$

where the "rest mass" m_0 represents the mass of a body that is not moving and c is the speed of light, which is about $3 \times 10^5 \, \text{km} \cdot \text{sec}^{-1}$ or about $186{,}000 \, \text{mi} \cdot \text{sec}^{-1}$.

For those who want to learn just enough about it so they can solve problems, that is all there is to the theory of relativity—it just changes Newton's laws by introducing a correction factor to the mass. From the formula itself it is easy to see that this mass increase is very small in ordinary circumstances. If the velocity is even as great as that of a satellite, which goes around the earth at 5 mi/sec, then $v/c = 5/186,000$: putting this value into the formula shows that the correction to the mass is only one part in two to three billion, which is nearly impossible to observe. Actually, the correctness of the formula has been amply confirmed by the observation of many kinds of particles, moving at speeds ranging up to practically the speed of light. However, because the effect is ordinarily so small, it seems remarkable that it was discovered theoretically before it was discovered experimentally. Empirically, at a sufficiently high velocity, the effect is very large, but it was not discovered that way. Therefore it is interesting to see how a law that involved so delicate a modification (at the time when it was first discovered) was brought to light by a combination of experiments and physical reasoning. Contributions to the discovery were made by a number of people, the final result of whose work was Einstein's discovery.

There are really two Einstein theories of relativity. This chapter is concerned with the Special Theory of Relativity, which dates from 1905. In 1915 Einstein published an additional theory, called the General Theory of Relativity. This latter theory deals with the extension of the Special Theory to the case of the law of gravitation; we shall not discuss the General Theory here.

The principle of relativity was first stated by Newton, in one of his corollaries to the laws of motion: "The motions of bodies included in a given space are the same among themselves, whether that space is at rest or moves uniformly forward in a straight line." This means, for example, that if a space ship is drifting along at a uniform speed, all experiments performed in the space ship and all the phenomena in the space ship will appear the same as if the ship were not moving, provided, of course, that one does not look outside. That is the meaning of the principle of relativity. This is a simple enough idea, and the only question is whether it is *true* that in all experiments performed inside a moving system the laws of physics will appear the same as they would if the system were

The Special Theory of Relativity

standing still. Let us first investigate whether Newton's laws appear the same in the moving system.

Suppose that Moe is moving in the x-direction with a uniform velocity u, and he measures the position of a certain point, shown in Figure 3-1. He designates the "x-distance" of the point in his coordinate system as x'. Joe is at rest, and measures the position of the same point, designating its x-coordinate in his system as x. The relationship of the coordinates in the two systems is clear from the diagram. After time t Moe's origin has moved a distance ut, and if the two systems originally coincided,

$$
\begin{aligned}
x' &= x - ut, \\
y' &= y, \\
z' &= z, \\
t' &= t.
\end{aligned}
\tag{3.2}
$$

If we substitute this transformation of coordinates into Newton's laws we find that these laws transform to the same laws in the primed system; that is, the laws of Newton are of the same form in a moving system as in a stationary system, and therefore it is impossible to tell, by making mechanical experiments, whether the system is moving or not.

The principle of relativity has been used in mechanics for a long time. It was employed by various people, in particular Huygens, to obtain the rules for the collision of billiard balls, in much the same

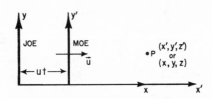

Figure 3-1. Two coordinate systems in uniform relative motion along their x-axes.

Six Not-So-Easy Pieces

way as we used it in Chapter 10* to discuss the conservation of momentum. In the past century interest in it was heightened as the result of investigations into the phenomena of electricity, magnetism, and light. A long series of careful studies of these phenomena by many people culminated in Maxwell's equations of the electromagnetic field, which describe electricity, magnetism, and light in one uniform system. However, the Maxwell equations did *not* seem to obey the principle of relativity. That is, if we transform Maxwell's equations by the substitution of equations 3.2, *their form does not remain the same*; therefore, in a moving space ship the electrical and optical phenomena should be different from those in a stationary ship. Thus one could use these optical phenomena to determine the speed of the ship; in particular, one could determine the absolute speed of the ship by making suitable optical or electrical measurements. One of the consequences of Maxwell's equations is that if there is a disturbance in the field such that light is generated, these electromagnetic waves go out in all directions equally and at the same speed c, or 186,000 mi/sec. Another consequence of the equations is that if the source of the disturbance is moving, the light emitted goes through space at the same speed c. This is analogous to the case of sound, the speed of sound waves being likewise independent of the motion of the source.

This independence of the motion of the source, in the case of light, brings up an interesting problem:

Suppose we are riding in a car that is going at a speed u, and light from the rear is going past the car with speed c. Differentiating the first equation in (3.2) gives

$$dx'/dt = dx/dt - u,$$

which means that according to the Galilean transformation the apparent speed of the passing light, as we measure it in the car, should not be c but should be $c - u$. For instance, if the car is going

* of the original *Lectures on Physics* vol. I.

The Special Theory of Relativity

100,000 mi/sec, and the light is going 186,000 mi/sec, then apparently the light going past the car should go 86,000 mi/sec. In any case, by measuring the speed of the light going past the car (if the Galilean transformation is correct for light), one could determine the speed of the car. A number of experiments based on this general idea were performed to determine the velocity of the earth, but they all failed—they gave *no velocity at all*. We shall discuss one of these experiments in detail, to show exactly what was done and what was the matter; something *was* the matter, of course, something was wrong with the equations of physics. What could it be?

3-2 *The Lorentz transformation*

When the failure of the equations of physics in the above case came to light, the first thought that occurred was that the trouble must lie in the new Maxwell equations of electrodynamics, which were only 20 years old at the time. It seemed almost obvious that these equations must be wrong, so the thing to do was to change them in such a way that under the Galilean transformation the principle of relativity would be satisfied. When this was tried, the new terms that had to be put into the equations led to predictions of new electrical phenomena that did not exist at all when tested experimentally, so this attempt had to be abandoned. Then it gradually became apparent that Maxwell's laws of electrodynamics were correct, and the trouble must be sought elsewhere.

In the meantime, H. A. Lorentz noticed a remarkable and curious thing when he made the following substitutions in the Maxwell equations:

$$x' = \frac{x - ut}{\sqrt{1 - u^2/c^2}},$$
$$y' = y,$$
$$z' = z, \qquad\qquad (3.3)$$
$$t' = \frac{t - ux/c^2}{\sqrt{1 - u^2/c^2}},$$

namely, Maxwell's equations remain in the same form when this transformation is applied to them! Equations (3.3) are known as a *Lorentz transformation*. Einstein, following a suggestion originally made by Poincaré, then proposed that *all the physical laws* should be of such a kind that they *remain unchanged under a Lorentz transformation*. In other words, we should change, not the laws of electrodynamics, but the laws of mechanics. How shall we change Newton's laws so that *they* will remain unchanged by the Lorentz transformation? If this goal is set, we then have to rewrite Newton's equations in such a way that the conditions we have imposed are satisfied. As it turned out, the only requirement is that the mass m in Newton's equations must be replaced by the form shown in Eq. (3.1). When this change is made, Newton's laws and the laws of electrodynamics will harmonize. Then if we use the Lorentz transformation in comparing Moe's measurements with Joe's, we shall never be able to detect whether either is moving, because the form of all the equations will be the same in both coordinate systems!

It is interesting to discuss what it means that we replace the old transformation between the coordinates and time with a new one, because the old one (Galilean) seems to be self-evident, and the new one (Lorentz) looks peculiar. We wish to know whether it is logically and experimentally possible that the new, and not the old, transformation can be correct. To find that out, it is not enough to study the laws of mechanics but, as Einstein did, we too must analyze our ideas of *space* and *time* in order to understand this transformation. We shall have to discuss these ideas and their implications for mechanics at some length, so we say in advance that the effort will be justified, since the results agree with experiment.

3-3 The Michelson-Morley experiment

As mentioned above, attempts were made to determine the absolute velocity of the earth through the hypothetical "ether" that was supposed to pervade all space. The most famous of these experiments is

The Special Theory of Relativity

one performed by Michelson and Morley in 1887. It was 18 years later before the negative results of the experiment were finally explained, by Einstein.

The Michelson-Morley experiment was performed with an apparatus like that shown schematically in Figure 3-2. This apparatus is essentially comprised of a light source A, a partially silvered glass plate B, and two mirrors C and E, all mounted on a rigid base. The mirrors are placed at equal distances L from B. The plate B splits an oncoming beam of light, and the two resulting beams continue in mutually perpendicular directions to the mirrors, where they are reflected back to B. On arriving back at B, the two beams are recombined as two superposed beams, D and F. If the time taken for the light to go from B to E and back is the same as the time from B to C and back, the emerging beams D and F will be in phase and will reinforce each other, but if the two times differ slightly, the beams will be slightly out of phase and interference will result. If the apparatus is "at rest" in the ether, the times should be precisely equal, but if it is moving toward the right with a velocity u, there should be a difference in the times. Let us see why.

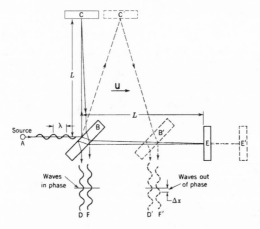

Figure 3-2. Schematic diagram of the Michelson-Morley experiment.

First, let us calculate the time required for the light to go from B to E and back. Let us say that the time for light to go from plate B to mirror E is t_1, and the time for the return is t_2. Now, while the light is on its way from B to the mirror, the apparatus moves a distance ut_1, so the light must traverse a distance $L + ut_1$, at the speed c. We can also express this distance as ct_1, so we have

$$ct_1 = L + ut_1, \quad \text{or} \quad t_1 = L/(c - u).$$

(This result is also obvious from the point of view that the velocity of light relative to the apparatus is $c - u$, so the time is the length L divided by $c - u$.) In a like manner, the time t_2 can be calculated. During this time the plate B advances a distance ut_2, so the return distance of the light is $L - ut_2$. Then we have

$$ct_2 = L - ut_2, \quad \text{or} \quad t_2 = L/(c + u).$$

Then the total time is

$$t_1 + t_2 = 2Lc/(c^2 - u^2).$$

For convenience in later comparison of times we write this as

$$t_1 + t_2 = \frac{2L/c}{1 - u^2/c^2}. \tag{3.4}$$

Our second calculation will be of the time t_3 for the light to go from B to the mirror C. As before, during time t_3 the mirror C moves to the right a distance ut_3 to the position C'; in the same time, the light travels a distance ct_3 along the hypotenuse of a triangle, which is BC'. For this right triangle we have

$$(ct_3)^2 = L^2 + (ut_3)^2$$

or

$$L^2 = c^2 t_3^2 - u^2 t_3^2 = (c^2 - u^2) t_3^2,$$

from which we get

$$t_3 = L/\sqrt{c^2 - u^2}.$$

The Special Theory of Relativity

For the return trip from C' the distance is the same, as can be seen from the symmetry of the figure; therefore the return time is also the same, and the total time is $2t_3$. With a little rearrangement of the form we can write

$$2t_3 = \frac{2L}{\sqrt{c^2 - u^2}} = \frac{2L/c}{\sqrt{1 - u^2/c^2}} . \tag{3.5}$$

We are now able to compare the times taken by the two beams of light. In expressions (3.4) and (3.5) the numerators are identical, and represent the time that would be taken if the apparatus were at rest. In the denominators, the term u^2/c^2 will be small, unless u is comparable in size to c. The denominators represent the modifications in the times caused by the motion of the apparatus. And behold, these modifications are *not the same*—the time to go to C and back is a little less than the time to E and back, even though the mirrors are equidistant from B, and all we have to do is to measure that difference with precision.

Here a minor technical point arises—suppose the two lengths L are not exactly equal? In fact, we surely cannot make them exactly equal. In that case we simply turn the apparatus 90 degrees, so that BC is in the line of motion and BE is perpendicular to the motion. Any small difference in length then becomes unimportant, and what we look for is a *shift* in the interference fringes when we rotate the apparatus.

In carrying out the experiment, Michelson and Morley oriented the apparatus so that the line BE was nearly parallel to the earth's motion in its orbit (at certain times of the day and night). This orbital speed is about 18 miles per second, and any "ether drift" should be at least that much at some time of the day or night and at some time during the year. The apparatus was amply sensitive to observe such an effect, but no time difference was found—the velocity of the earth through the ether could not be detected. The result of the experiment was null.

The result of the Michelson-Morley experiment was very

puzzling and most disturbing. The first fruitful idea for finding a way out of the impasse came from Lorentz. He suggested that material bodies contract when they are moving, and that this foreshortening is only in the direction of the motion, and also, that if the length is L_0 when a body is at rest, then when it moves with speed u parallel to its length, the new length, which we call $L_\parallel$ (L-parallel), is given by

$$L_\parallel = L_0 \sqrt{1 - u^2/c^2}. \tag{3.6}$$

When this modification is applied to the Michelson-Morley interferometer apparatus the distance from B to C does not change, but the distance from B to E is shortened to $L\sqrt{1 - u^2/c^2}$. Therefore Eq. (3.5) is not changed, but the L of Eq. (3.4) must be changed in accordance with Eq. (3.6). When this is done we obtain

$$t_1 + t_2 = \frac{(2L/c)\sqrt{1 - u^2/c^2}}{1 - u^2/c^2} = \frac{2L/c}{\sqrt{1 - u^2/c^2}}. \tag{3.7}$$

Comparing this result with Eq. (3.5), we see that $t_1 + t_2 = 2t_3$. So if the apparatus shrinks in the manner just described, we have a way of understanding why the Michelson-Morley experiment gives no effect at all. Although the contraction hypothesis successfully accounted for the negative result of the experiment, it was open to the objection that it was invented for the express purpose of explaining away the difficulty, and was too artificial. However, in many other experiments to discover an ether wind, similar difficulties arose, until it appeared that nature was in a "conspiracy" to thwart man by introducing some new phenomenon to undo every phenomenon that he thought would permit a measurement of u.

It was ultimately recognized, as Poincaré pointed out, that *a complete conspiracy is itself a law of nature*! Poincaré then proposed that there *is* such a law of nature, that it is not possible to discover an ether wind by *any* experiment; that is, there is no way to determine an absolute velocity.

The Special Theory of Relativity

3-4 Transformation of time

In checking out whether the contraction idea is in harmony with the facts in other experiments, it turns out that everything is correct provided that the *times* are also modified, in the manner expressed in the fourth equation of the set (3.3). That is because the time t_3, calculated for the trip from B to C and back, is not the same when calculated by a man performing the experiment in a moving space ship as when calculated by a stationary observer who is watching the space ship. To the man in the ship the time is simply $2L/c$, but to the other observer it is $(2L/c)/\sqrt{1 - u^2/c^2}$ (Eq. 3.5). In other words, when the outsider sees the man in the space ship lighting a cigar, all the actions appear to be slower than normal, while to the man inside, everything moves at a normal rate. So not only must the lengths shorten, but also the time-measuring instruments ("clocks") must apparently slow down. That is, when the clock in the space ship records 1 second elapsed, as seen by the man in the ship, it shows $1/\sqrt{1 - u^2/c^2}$ second to the man outside.

This slowing of the clocks in a moving system is a very peculiar phenomenon, and is worth an explanation. In order to understand this, we have to watch the machinery of the clock and see what happens when it is moving. Since that is rather difficult, we shall take a very simple kind of clock. The one we choose is rather a silly kind of clock, but it will work in principle: it is a rod (meter stick) with a mirror at each end, and when we start a light signal between the mirrors, the light keeps going up and down, making a click every time it comes down, like a standard ticking clock. We build two such clocks, with exactly the same lengths, and synchronize them by starting them together; then they agree always thereafter, because they are the same in length, and light always travels with speed c. We give one of these clocks to the man to take along in his space ship, and he mounts the rod perpendicular to the direction of motion of the ship; then the length of the rod will not change. How do we know that perpendicular lengths do not change? The men

SIX NOT-SO-EASY PIECES

can agree to make marks on each other's *y*-meter stick as they pass each other. By symmetry, the two marks must come at the same *y*- and *y'*-coordinates, since otherwise, when they get together to compare results, one mark will be above or below the other, and so we could tell who was really moving.

Now let us see what happens to the moving clock. Before the man took it aboard, he agreed that it was a nice, standard clock, and when he goes along in the space ship he will not see anything peculiar. If he did, he would know he was moving—if anything at all changed because of the motion, he could tell he was moving. But the principle of relativity says this is impossible in a uniformly moving system, so nothing has changed. On the other hand, when the external observer looks at the clock going by, he sees that the light, in going from mirror to mirror, is "really" taking a zigzag path, since the rod is moving sideways all the while. We have already analyzed such a zigzag motion in connection with the Michelson-Morley experiment. If in a given time the rod moves forward a distance proportional to *u* in Figure 3-3, the distance the light

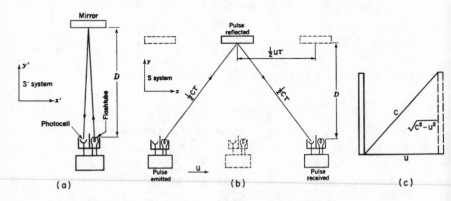

Figure 3-3. (a) A "light clock" at rest in the *S'* system. (b) The same clock, moving through the *S* system. (c) Illustration of the diagonal path taken by the light beam in a moving "light clock."

The Special Theory of Relativity

travels in the same time is proportional to c, and the vertical distance is therefore proportional to $\sqrt{c^2 - u^2}$.

That is, it takes a *longer time* for light to go from end to end in the moving clock than in the stationary clock. Therefore the apparent time between clicks is longer for the moving clock, in the same proportion as shown in the hypotenuse of the triangle (that is the source of the square root expressions in our equations). From the figure it is also apparent that the greater u is, the more slowly the moving clock appears to run. Not only does this particular kind of clock run more slowly, but if the theory of relativity is correct, any other clock, operating on any principle whatsoever, would also appear to run slower, and in the same proportion—we can say this without further analysis. Why is this so?

To answer the above question, suppose we had two other clocks made exactly alike with wheels and gears, or perhaps based on radioactive decay, or something else. Then we adjust these clocks so they both run in precise synchronism with our first clocks. When light goes up and back in the first clocks and announces its arrival with a click, the new models also complete some sort of cycle, which they simultaneously announce by some doubly coincident flash, or bong, or other signal. One of these clocks is taken into the space ship, along with the first kind. Perhaps *this* clock will not run slower, but will continue to keep the same time as its stationary counterpart, and thus disagree with the other moving clock. Ah no, if that should happen, the man in the ship could use this mismatch between his two clocks to determine the speed of his ship, which we have been supposing is impossible. *We need not know anything about the machinery* of the new clock that might cause the effect—we simply know that whatever the reason, it will appear to run slow, just like the first one.

Now if *all* moving clocks run slower, if no way of measuring time gives anything but a slower rate, we shall just have to say, in a certain sense, that *time itself* appears to be slower in a space ship. All the phenomena there—the man's pulse rate, his thought processes, the time he takes to light a cigar, how long it takes to grow up and get old—all these things must be slowed down in the same proportion, because he cannot tell he is moving. The biologists and medical men sometimes say it is not quite certain that the time it takes for a cancer to develop will be longer in a space ship, but from the viewpoint of a modern physicist it is nearly certain; otherwise one could use the rate of cancer development to determine the speed of the ship!

A very interesting example of the slowing of time with motion is furnished by mu-mesons (muons), which are particles that disintegrate spontaneously after an average lifetime of 2.2×10^{-6} sec. They come to the earth in cosmic rays, and can also be produced artificially in the laboratory. Some of them disintegrate in midair, but the remainder disintegrate only after they encounter a piece of material and stop. It is clear that in its short lifetime a muon cannot travel, even at the speed of light, much more than 600 meters. But although the muons are created at the top of the atmosphere, some 10 kilometers up, yet they are actually found in a laboratory down here, in cosmic rays. How can that be? The answer is that different muons move at various speeds, some of which are very close to the speed of light. While from their own point of view they live only about 2 μsec, from our point of view they live considerably longer—enough longer that they may reach the earth. The factor by which the time is increased has already been given as $1/\sqrt{1 - u^2/c^2}$. The average life has been measured quite accurately for muons of different velocities, and the values agree closely with the formula.

We do not know why the meson disintegrates or what its machinery is, but we do know its behavior satisfies the principle of relativity. That is the utility of the principle of relativity—it permits us to make predictions, even about things that otherwise we do not know much about. For example, before we have any idea at all about what makes the meson disintegrate, we can still predict that when it is moving at nine-tenths of the speed of light, the apparent length of time that it lasts is $(2.2 \times 10^{-6})/\sqrt{1 - 9^2/10^2}$ sec; and our prediction works—that is the good thing about it.

The Special Theory of Relativity

3-5 The Lorentz contraction

Now let us return to the Lorentz transformation (3.3) and try to get a better understanding of the relationship between the (x, y, z, t) and the (x', y', z', t') coordinate systems, which we shall call the S and S' systems, or Joe and Moe systems, respectively. We have already noted that the first equation is based on the Lorentz suggestion of contraction along the x-direction; how can we prove that a contraction takes place? In the Michelson-Morley experiment, we now appreciate that the *transverse* arm BC cannot change length, by the principle of relativity; yet the null result of the experiment demands that the *times* must be equal. So, in order for the experiment to give a null result, the longitudinal arm BE must appear shorter, by the square root $\sqrt{1 - u^2/c^2}$. What does this contraction mean, in terms of measurements made by Joe and Moe? Suppose that Moe, moving with the S' system in the x-direction, is measuring the x'-coordinate of some point with a meter stick. He lays the stick down x' times, so he thinks the distance is x' meters. From the viewpoint of Joe in the S system, however, Moe is using a foreshortened ruler, so the "real" distance measured is $x'\sqrt{1 - u^2/c^2}$ meters. Then if the S' system has travelled a distance ut away from the S system, the S observer would say that the same point, measured in his coordinates, is at a distance $x = x'\sqrt{1 - u^2/c^2} + ut$, or

$$x' = \frac{x - ut}{\sqrt{1 - u^2/c^2}},$$

which is the first equation of the Lorentz transformation.

3-6 Simultaneity

In an analogous way, because of the difference in time scales, the denominator expression is introduced into the fourth equation of the Lorentz transformation. The most interesting term in that equation is the ux/c^2 in the numerator, because that is quite new and unexpected. Now what does that mean? If we look at the situation carefully we see that events that occur at two separated places at the same time, as seen by Moe in S', do *not* happen at the same time as viewed by Joe in S. If one event occurs at point x_1 at time t_0 and the other event at x_2 and t_0 (the same time), we find that the two corresponding times t'_1 and t'_2 differ by an amount

$$t'_2 - t'_1 = \frac{u(x_1 - x_2)/c^2}{\sqrt{1 - u^2/c^2}} .$$

This circumstance is called "failure of simultaneity at a distance," and to make the idea a little clearer let us consider the following experiment.

Suppose that a man moving in a space ship (system S') has placed a clock at each end of the ship and is interested in making sure that the two clocks are in synchronism. How can the clocks be synchronized? There are many ways. One way, involving very little calculation, would be first to locate exactly the midpoint between the clocks. Then from this station we send out a light signal which will go both ways at the same speed and will arrive at both clocks, clearly, at the same time. This simultaneous arrival of the signals can be used to synchronize the clocks. Let us then suppose that the man in S' synchronizes his clocks by this particular method. Let us see whether an observer in system S would agree that the two clocks are synchronous. The man in S' has a right to believe they are, because he does not know that he is moving. But the man in S reasons that since the ship is moving forward, the clock in the front end was running away from the light signal, hence the light had to go more than halfway in order to catch up; the rear clock, however, was advancing to meet the light signal, so this distance was shorter. Therefore the signal reached the rear clock first, although the man in S' thought that the signals arrived simultaneously. We thus see that when a man in a space ship thinks the times at two locations are simultaneous, equal values of t' in his coordinate system must correspond to *different* values of t in the other coordinate system!

The Special Theory of Relativity

3-7 Four-vectors

Let us see what else we can discover in the Lorentz transformation. It is interesting to note that the transformation between the x's and t's is analogous in form to the transformation of the x's and y's that we studied in Chapter 1 for a rotation of coordinates. We then had

$$x' = x \cos \theta + y \sin \theta,$$
$$y' = y \cos \theta - x \sin \theta, \qquad (3.8)$$

in which the new x' mixes the old x and y, and the new y' also mixes the old x and y; similarly, in the Lorentz transformation we find a new x' which is a mixture of x and t, and a new t' which is a mixture of t and x. So the Lorentz transformation is analogous to a rotation, only it is a "rotation" in *space and time*, which appears to be a strange concept. A check of the analogy to rotation can be made by calculating the quantity

$$x'^2 + y'^2 + z'^2 - c^2t'^2 = x^2 + y^2 + z^2 - c^2t^2. \qquad (3.9)$$

In this equation the first three terms on each side represent, in three-dimensional geometry, the square of the distance between a point and the origin (surface of a sphere) which remains unchanged (invariant) regardless of rotation of the coordinate axes. Similarly, Eq. (3.9) shows that there is a certain combination which includes time, that is invariant to a Lorentz transformation. Thus, the analogy to a rotation is complete, and is of such a kind that vectors, i.e., quantities involving "components" which transform the same way as the coordinates and time, are also useful in connection with relativity.

Thus we contemplate an extension of the idea of vectors, which we have so far considered to have only space components, to include a time component. That is, we expect that there will be vectors with four components, three of which are like the components of an ordinary vector, and with these will be associated a fourth component, which is the analog of the time part.

This concept will be analyzed further in the next chapters, where we shall find that if the ideas of the preceding paragraph are applied to momentum, the transformation gives three space parts that are like ordinary momentum components, and a fourth component, the time part, which is the *energy*.

3-8 Relativistic dynamics

We are now ready to investigate, more generally, what form the laws of mechanics take under the Lorentz transformation. [We have thus far explained how length and time change, but not how we get the modified formula for m (Eq. 3.1). We shall do this in the next chapter.] To see the consequences of Einstein's modification of m for Newtonian mechanics, we start with the Newtonian law that force is the rate of change of momentum, or

$$\mathbf{F} = d(m\mathbf{v})/dt.$$

Momentum is still given by mv, but when we use the new m this becomes

$$\mathbf{p} = m\mathbf{v} = \frac{m_0 \mathbf{v}}{\sqrt{1 - v^2/c^2}}. \tag{3.10}$$

This is Einstein's modification of Newton's laws. Under this modification, if action and reaction are still equal (which they may not be in detail, but are in the long run), there will be conservation of momentum in the same way as before, but the quantity that is being conserved is not the old $m\mathbf{v}$ with its constant mass, but instead is the quantity shown in (3.10), which has the modified mass. When this change is made in the formula for momentum, conservation of momentum still works.

Now let us see how momentum varies with speed. In Newtonian mechanics it is proportional to the speed and, according to (3.10),

The Special Theory of Relativity

over a considerable range of speed, but small compared with c, it is nearly the same in relativistic mechanics, because the square-root expression differs only slightly from 1. But when v is almost equal to c, the square-root expression approaches zero, and the momentum therefore goes toward infinity.

What happens if a constant force acts on a body for a long time? In Newtonian mechanics the body keeps picking up speed until it goes faster than light. But this is impossible in relativistic mechanics. In relativity, the body keeps picking up, not speed, but momentum, which can continually increase because the mass is increasing. After a while there is practically no acceleration in the sense of a change of velocity, but the momentum continues to increase. Of course, whenever a force produces very little change in the velocity of a body, we say that the body has a great deal of inertia, and that is exactly what our formula for relativistic mass says (see Eq. 3.10)—it says that the inertia is very great when v is nearly as great as c. As an example of this effect, to deflect the high-speed electrons in the synchrotron that is used here at Caltech, we need a magnetic field that is 2000 times stronger than would be expected on the basis of Newton's laws. In other words, the mass of the electrons in the synchrotron is 2000 times as great as their normal mass, and is as great as that of a proton! That m should be 2000 times m_0 means that $1 - v^2/c^2$ must be 1/4,000,000, and that means that v^2/c^2 differs from 1 by one part in 4,000,000, or that v differs from c by one part in 8,000,000, so the electrons are getting pretty close to the speed of light. If the electrons and light were both to start from the synchrotron (estimated as 700 feet away) and rush out to Bridge Lab, which would arrive first? The light, of course, because light always travels faster.* How much earlier? That is too hard to tell—instead, we tell by what distance the light is ahead: it is about 1/1000 of an inch, or ¼ the thickness of a piece of paper!

---

* The electrons would actually win the race versus *visible* light because of the index of refraction of air. A gamma ray would make out better.

When the electrons are going that fast their masses are enormous, but their speed cannot exceed the speed of light.

Now let us look at some further consequences of relativistic change of mass. Consider the motion of the molecules in a small tank of gas. When the gas is heated, the speed of the molecules is increased, and therefore the mass is also increased and the gas is heavier. An approximate formula to express the increase of mass, for the case when the velocity is small, can be found by expanding $m_0/\sqrt{1 - v^2/c^2} = m_0(1 - v^2/c^2)^{-1/2}$ in a power series, using the binomial theorem. We get

$$m_0(1 - v^2/c^2)^{-1/2} = m_0(1 + \tfrac{1}{2}v^2/c^2 + \tfrac{3}{8}v^4/c^4 + \cdots).$$

We see clearly from the formula that the series converges rapidly when v is small, and the terms after the first two or three are negligible. So we can write

$$m \cong m_0 + \tfrac{1}{2}m_0 v^2 \left(\frac{1}{c^2}\right) \tag{3.11}$$

in which the second term on the right expresses the increase of mass due to molecular velocity. When the temperature increases the v^2 increases proportionately, so we can say that the increase in mass is proportional to the increase in temperature. But since $\tfrac{1}{2}\,m_0 v^2$ is the kinetic energy in the old-fashioned Newtonian sense, we can also say that the increase in mass of all this body of gas is equal to the increase in kinetic energy divided by c^2, or $\Delta m = \Delta(\text{K.E.})/c^2$.

3-9 Equivalence of mass and energy

The above observation led Einstein to the suggestion that the mass of a body can be expressed more simply than by the formula (3.1), if we say that the mass is equal to the total energy content divided by c^2. If Eq. (3.11) is multiplied by c^2 the result is

$$mc^2 = m_0 c^2 + \tfrac{1}{2}m_0 v^2 + \cdots. \tag{3.12}$$

Here, the term on the left expresses the total energy of a body, and we recognize the last term as the ordinary kinetic energy. Einstein

The Special Theory of Relativity

interpreted the large constant term, m_0c^2, to be part of the total energy of the body, an intrinsic energy known as the "rest energy."

Let us follow out the consequences of assuming, with Einstein, that *the energy of a body always equals* mc^2. As an interesting result, we shall find the formula (3.1) for the variation of mass with speed, which we have merely assumed up to now. We start with the body at rest, when its energy is m_0c^2. Then we apply a force to the body, which starts it moving and gives it kinetic energy; therefore, since the energy has increased, the mass has increased—this is implicit in the original assumption. So long as the force continues, the energy and the mass both continue to increase. We have already seen (Chapter 13*) that the rate of change of energy with time equals the force times the velocity, or

$$\frac{dE}{dt} = \mathbf{F} \cdot \mathbf{v}.$$

(3.13)

We also have (Chapter 9*, Eq. 9.1) that $F = d(mv)/dt$. When these relations are put together with the definition of E, Eq. (3.13) becomes

$$\frac{d(mc^2)}{dt} = \mathbf{v} \cdot \frac{d(m\mathbf{v})}{dt}.$$

(3.14)

We wish to solve this equation for m. To do this we first use the mathematical trick of multiplying both sides by $2m$, which changes the equation to

$$c^2(2m) \frac{dm}{dt} = 2mv \frac{d(mv)}{dt}.$$

(3.15)

We need to get rid of the derivatives, which can be accomplished by integrating both sides. The quantity $(2m)\, dm/dt$ can be recognized

---

* of the original *Lectures on Physics*, vol. I.

as the time derivative of m^2, and $(2m\mathbf{v}) \cdot d(m\mathbf{v})/dt$ is the time derivative of $(mv)^2$. So, Eq. (3.15) is the same as

$$c^2 \frac{d(m^2)}{dt} = \frac{d(m^2v^2)}{dt}. \tag{3.16}$$

If the derivatives of two quantities are equal, the quantities themselves differ at most by a constant, say C. This permits us to write

$$m^2c^2 = m^2v^2 + C. \tag{3.17}$$

We need to define the constant C more explicitly. Since Eq. (3.17) must be true for all velocities, we can choose a special case where $v = 0$, and say that in this case the mass is m_0. Substituting these values into Eq. (3.17) gives

$$m_0^2c^2 = 0 + C.$$

We can now use this value of C in Eq. (3.17), which becomes

$$m^2c^2 = m^2v^2 + m_0^2c^2. \tag{3.18}$$

Dividing by c^2 and rearranging terms gives

$$m^2(1 - v^2/c^2) = m_0^2,$$

from which we get

$$m = m_0/\sqrt{1 - v^2/c^2}. \tag{3.19}$$

This is the formula (3.1), and is exactly what is necessary for the agreement between mass and energy in Eq. (3.12).

Ordinarily these energy changes represent extremely slight changes in mass, because most of the time we cannot generate much energy from a given amount of material; but in an atomic bomb of explosive energy equivalent to 20 kilotons of TNT, for example, it can be shown that the dirt after the explosion is lighter by 1 gram than the initial mass of the reacting material, because of the energy that was released, i.e., the released energy had a mass of 1 gram, according to the relationship $\Delta E = \Delta(mc^2)$. This theory of equivalence of mass and energy has been beautifully verified by experi-

The Special Theory of Relativity

ments in which matter is annihilated—converted totally to energy: An electron and a positron come together at rest, each with a rest mass m_0. When they come together they disintegrate and two gamma rays emerge, each with the measured energy of $m_0 c^2$. This experiment furnishes a direct determination of the energy associated with the existence of the rest mass of a particle.

Four

RELATIVISTIC ENERGY AND MOMENTUM

4-1 Relativity and the philosophers

▶ In this chapter we shall continue to discuss the principle of relativity of Einstein and Poincaré, as it affects our ideas of physics and other branches of human thought.

Poincaré made the following statement of the principle of relativity: "According to the principle of relativity, the laws of physical phenomena must be the same for a fixed observer as for an observer who has a uniform motion of translation relative to him, so that we have not, nor can we possibly have, any means of discerning whether or not we are carried along in such a motion."

When this idea descended upon the world, it caused a great stir among philosophers, particularly the "cocktail-party philosophers," who say, "Oh, it is very simple: Einstein's theory says all is relative!" In fact, a surprisingly large number of philosophers, not only those found at cocktail parties (but rather than embarrass them, we shall just call them "cocktail-party philosophers"), will say, "That all is relative is a consequence of Einstein, and it has profound influences on our ideas." In addition, they say "It has been demonstrated in physics that phenomena depend upon your frame of reference." We hear that a great deal, but it is difficult to find out what it means. Probably the frames of reference that were originally referred to were the coordinate systems which we use in the analysis of the theory of relativity. So the fact that "things depend upon your frame of reference" is supposed to have had a profound effect on modern thought. One might well wonder why, because, after all, that things depend upon one's point of view is so simple an idea that it certainly cannot have been necessary to go to all the trouble of the physical relativity theory in order to discover it. That what one sees depends upon his frame of reference is certainly known to anybody who walks around, because he sees an approaching pedestrian first from the front and then from the back; there is nothing deeper in most of the philosophy which is said to have come from the theory of relativity than the remark that "A person looks different from the front than from the back." The old story about the elephant that several blind men describe in different ways is another example, perhaps, of the theory of relativity from the philosopher's point of view.

But certainly there must be deeper things in the theory of relativity than just this simple remark that "A person looks different from the front than from the back." Of course relativity is deeper than this, because *we can make definite predictions with it*. It certainly would be rather remarkable if we could predict the behavior of nature from such a simple observation alone.

There is another school of philosophers who feel very uncomfortable about the theory of relativity, which asserts that we cannot determine our absolute velocity without looking at something outside, and who would say, "It is obvious that one cannot measure his velocity without looking outside. It is self-evident that it is *meaningless* to talk about the velocity of a thing without looking outside; the physicists are rather stupid for having thought otherwise, but it has just dawned on them that this is the case. If only we philosophers had realized what the problems were that the physicists had, we could have decided immediately by brainwork that it is impossible to tell how fast one is moving without looking outside, and we could have made an enormous contribution to physics." These philosophers are always with us, struggling in the periphery to try to tell us

Relativistic Energy and Momentum

something, but they never really understand the subtleties and depths of the problem.

Our inability to detect absolute motion is a result of *experiment* and not a result of plain thought, as we can easily illustrate. In the first place, Newton believed that it was true that one could not tell how fast he is going if he is moving with uniform velocity in a straight line. In fact, Newton first stated the principle of relativity, and one quotation made in the last chapter was a statement of Newton's. Why then did the philosophers not make all this fuss about "all is relative," or whatever, in Newton's time? Because it was not until Maxwell's theory of electrodynamics was developed that there were physical laws that suggested that one *could* measure his velocity without looking outside; soon it was found *experimentally* that one could *not*.

Now, *is* it absolutely, definitely, philosophically *necessary* that one should not be able to tell how fast he is moving without looking outside? One of the consequences of relativity was the development of a philosophy which said, "You can only define what you can measure! Since it is self-evident that one cannot measure a velocity without seeing what he is measuring it relative to, therefore it is clear that there is no *meaning* to absolute velocity. The physicists should have realized that they can talk only about what they can measure." But *that is the whole problem:* whether or not one *can define* absolute velocity is the same as the problem of whether or not one *can detect in an experiment*, without looking outside, whether he is moving. In other words, whether or not a thing is measurable is not something to be decided *a priori* by thought alone, but something that can be decided only by experiment. Given the fact that the velocity of light is 186,000 mi/sec, one will find few philosophers who will calmly state that it is self-evident that if light goes 186,000 mi/sec inside a car, and the car is going 100,000 mi/sec, that the light also goes 186,000 mi/sec past an observer on the ground. That is a shocking fact to them; the very ones who claim it is obvious find, when you give them a specific fact, that it is not obvious.

Finally, there is even a philosophy which says that one cannot detect *any* motion except by looking outside. It is simply not true in physics. True, one cannot perceive a *uniform* motion in a *straight line*, but if the whole room were *rotating* we would certainly know it, for everybody would be thrown to the wall—there would be all kinds of "centrifugal" effects. That the earth is turning on its axis can be determined without looking at the stars, by means of the so-called Foucault pendulum, for example. Therefore it is not true that "all is relative"; it is only *uniform velocity* that cannot be detected without looking outside. Uniform *rotation* about a fixed axis *can* be. When this is told to a philosopher, he is very upset that he did not really understand it, because to him it seems impossible that one should be able to determine rotation about an axis without looking outside. If the philosopher is good enough, after some time he may come back and say, "I understand. We really do not have such a thing as absolute rotation; we are really rotating *relative to the stars*, you see. And so some influence exerted by the stars on the object must cause the centrifugal force."

Now, for all we know, that is true; we have no way, at the present time, of telling whether there would have been centrifugal force if there were no stars and nebulae around. We have not been able to do the experiment of removing all the nebulae and then measuring our rotation, so we simply do not know. We must admit that the philosopher may be right. He comes back, therefore, in delight and says, "It is absolutely necessary that the world ultimately turn out to be this way: *absolute* rotation means nothing; it is only *relative* to the nebulae." Then we say to him, "*Now*, my friend, is it or is it not obvious that uniform velocity in a straight line, *relative to the nebulae*, should produce no effects inside a car?" Now that the motion is no longer absolute, but is a motion *relative to the nebulae*, it becomes a mysterious question, and a question that can be answered only by experiment.

What, then, *are* the philosophic influences of the theory of relativity? If we limit ourselves to influences in the sense of *what kind of new ideas and suggestions* are made to the physicist by the principle of relativity, we could describe some of them as follows. The first

Relativistic Energy and Momentum

discovery is, essentially, that even those ideas which have been held for a very long time and which have been very accurately verified might be wrong. It was a shocking discovery, of course, that Newton's laws are wrong, after all the years in which they seemed to be accurate. Of course it is clear, not that the experiments were wrong, but that they were done over only a limited range of velocities, so small that the relativistic effects would not have been evident. But nevertheless, we now have a much more humble point of view of our physical laws—everything *can* be wrong!

Secondly, if we have a set of "strange" ideas, such as that time goes slower when one moves, and so forth, whether we *like* them or do *not* like them is an irrelevant question. The only relevant question is whether the ideas are consistent with what is found experimentally. In other words, the "strange ideas" need only agree with *experiment*, and the only reason that we have to discuss the behavior of clocks and so forth is to demonstrate that although the notion of the time dilation is strange, it is *consistent* with the way we measure time.

Finally, there is a third suggestion which is a little more technical but which has turned out to be of enormous utility in our study of other physical laws, and that is to *look at the symmetry of the laws* or, more specifically, to look for the ways in which the laws can be transformed and leave their form the same. When we discussed the theory of vectors, we noted that the fundamental laws of motion are not changed when we rotate the coordinate system, and now we learn that they are not changed when we change the space and time variables in a particular way, given by the Lorentz transformation. So this idea of studying the patterns or operations under which the fundamental laws are not changed has proved to be a very useful one.

4-2 *The twin paradox*

To continue our discussion of the Lorentz transformation and relativistic effects, we consider a famous so-called "paradox" of Peter and Paul, who are supposed to be twins, born at the same time. When they are old enough to drive a space ship, Paul flies away at very high speed. Because Peter, who is left on the ground, sees Paul going so fast, all of Paul's clocks appear to go slower, his heart beats go slower, his thoughts go slower, everything goes slower, from Peter's point of view. Of course, Paul notices nothing unusual, but if he travels around and about for a while and then comes back, he will be younger than Peter, the man on the ground! That is actually right; it is one of the consequences of the theory of relativity which has been clearly demonstrated. Just as the mu-mesons last longer when they are moving, so also will Paul last longer when he is moving. This is called a "paradox" only by the people who believe that the principle of relativity means that *all motion* is relative; they say, "Heh, heh, heh, from the point of view of Paul, can't we say that *Peter* was moving and should therefore appear to age more slowly? By symmetry, the only possible result is that both should be the same age when they meet." But in order for them to come back together and make the comparison, Paul must either stop at the end of the trip and make a comparison of clocks or, more simply, he has to come back, and the one who comes back must be the man who was moving, and he knows this, because he had to turn around. When he turned around, all kinds of unusual things happened in his space ship—the rockets went off, things jammed up against one wall, and so on—while Peter felt nothing.

So the way to state the rule is to say that *the man who has felt the accelerations*, who has seen things fall against the walls, and so on, is the one who would be the younger; that is the difference between them in an "absolute" sense, and it is certainly correct. When we discussed the fact that moving mu-mesons live longer, we used as an example their straight-line motion in the atmosphere. But we can also make mu-mesons in a laboratory and cause them to go in a curve with a magnet, and even under this accelerated motion, they last exactly as much longer as they do when they are moving in a straight line. Although no one has arranged an experiment explic-itly so that we can get rid of the paradox, one could compare a mu-meson which is left standing with one that had gone around a

Relativistic Energy and Momentum

complete circle, and it would surely be found that the one that went around the circle lasted longer. Although we have not actually carried out an experiment using a complete circle, it is really not necessary, of course, because everything fits together all right. This may not satisfy those who insist that every single fact be demonstrated directly, but we confidently predict the result of the experiment in which Paul goes in a complete circle.

4-3 Transformation of velocities

The main difference between the relativity of Einstein and the relativity of Newton is that the laws of transformation connecting the coordinates and times between relatively moving systems are different. The correct transformation law, that of Lorentz, is

$$x' = \frac{x - ut}{\sqrt{1 - u^2/c^2}},$$
$$y' = y,$$
$$z' = z,$$
$$t' = \frac{t - ux/c^2}{\sqrt{1 - u^2/c^2}}.$$
$$(4.1)$$

These equations correspond to the relatively simple case in which the relative motion of the two observers is along their common x-axes. Of course other directions of motion are possible, but the most general Lorentz transformation is rather complicated, with all four quantities mixed up together. We shall continue to use this simpler form, since it contains all the essential features of relativity.

Let us now discuss more of the consequences of this transformation. First, it is interesting to solve these equations in reverse. That is, here is a set of linear equations, four equations with four unknowns, and they can be solved in reverse, for x, y, z, t in terms of x', y', z', t'. The result is very interesting, since it tells us how a system of coordinates "at rest" looks from the point of view of one that is "moving." Of course, since the motions are relative and of uniform velocity, the man who is "moving" can say, if he wishes, that

SIX NOT-SO-EASY PIECES

it is really the other fellow who is moving and he himself who is at rest. And since he is moving in the opposite direction, he should get the same transformation, but with the opposite sign of velocity. That is precisely what we find by manipulation, so that is consistent. If it did not come out that way, we would have real cause to worry!

$$x = \frac{x' + ut'}{\sqrt{1 - u^2/c^2}},$$
$$y = y',$$
$$z = z', \tag{4.2}$$
$$t = \frac{t' + ux'/c^2}{\sqrt{1 - u^2/c^2}}.$$

Next we discuss the interesting problem of the addition of velocities in relativity. We recall that one of the original puzzles was that light travels at 186,000 mi/sec in all systems, even when they are in relative motion. This is a special case of the more general problem exemplified by the following. Suppose that an object inside a space ship is going at 100,000 mi/sec and the space ship itself is going at 100,000 mi/sec; how fast is the object inside the space ship moving from the point of view of an observer outside? We might want to say 200,000 mi/sec, which is faster than the speed of light. This is very unnerving, because it is not supposed to be going faster than the speed of light! The general problem is as follows.

Let us suppose that the object inside the ship, from the point of view of the man inside, is moving with velocity v, and that the space ship itself has a velocity u with respect to the ground. We want to know with what velocity v_x this object is moving from the point of view of the man on the ground. This is, of course, still but a special case in which the motion is in the x-direction. There will also be a transformation for velocities in the y-direction, or for any angle; these can be worked out as needed. Inside the space ship the velocity is $v_{x'}$, which means that the displacement x is equal to the velocity times the time:

$$x' = v_{x'}t'. \tag{4.3}$$

Relativistic Energy and Momentum

Now we have only to calculate what the position and time are from the point of view of the outside observer for an object which has the relation (4.2) between x' and t'. So we simply substitute (4.3) into (4.2), and obtain

$$x = \frac{v_{x'}t' + ut'}{\sqrt{1 - u^2/c^2}} \,.$$ (4.4)

But here we find x expressed in terms of t'. In order to get the velocity as seen by the man on the outside, we must divide *his distance* by *his time*, not by the *other man's time*! So we must also calculate the *time* as seen from the outside, which is

$$t = \frac{t' + u(v_{x'}t')/c^2}{\sqrt{1 - u^2/c^2}} \,.$$ (4.5)

Now we must find the ratio of x to t, which is

$$v_x = \frac{x}{t} = \frac{u + v_{x'}}{1 + uv_{x'}/c^2} \,,$$ (4.6)

the square roots having cancelled. This is the law that we seek: the resultant velocity, the "summing" of two velocities, is not just the algebraic sum of two velocities (we know that it cannot be or we get in trouble), but is "corrected" by $1 + uv/c^2$.

Now let us see what happens. Suppose that you are moving inside the space ship at half the speed of light, and that the space ship itself is going at half the speed of light. Thus u is $\frac{1}{2}c$ and v is $\frac{1}{2}c$, but in the denominator uv is one-fourth, so that

$$v = \frac{\frac{1}{2}c + \frac{1}{2}c}{1 + \frac{1}{4}} = \frac{4c}{5} \,.$$

So, in relativity, "half" and "half" does not make "one," it makes only "4/5." Of course low velocities can be added quite easily in the familiar way, because so long as the velocities are small compared with the speed of light we can forget about the $(1 + uv/c^2)$ factor; but things are quite different and quite interesting at high velocity.

Let us take a limiting case. Just for fun, suppose that inside the space ship the man was observing *light itself*. In other words, $v = c$, and yet the space ship is moving. How will it look to the man on the ground? The answer will be

$$v = \frac{u + c}{1 + uc/c^2} = c\,\frac{u + c}{u + c} = c.$$

Therefore, if something is moving at the speed of light inside the ship, it will appear to be moving at the speed of light from the point of view of the man on the ground too! This is good, for it is, in fact, what the Einstein theory of relativity was designed to do in the first place—so it had *better* work!

Of course, there are cases in which the motion is not in the direction of the uniform translation. For example, there may be an object inside the ship which is just moving "upward" with the velocity $v_{y'}$ with respect to the ship, and the ship is moving "horizontally." Now, we simply go through the same thing, only using y's instead of x's, with the result

$$y = y' = v_{y'}t',$$

so that if $v_{x'} = 0$,

$$v_y = \frac{y}{t} = v_{y'}\,\sqrt{1 - u^2/c^2}. \tag{4.7}$$

Thus a sidewise velocity is no longer $v_{y'}$, but $v_{y'}\sqrt{1 - u^2/c^2}$. We found this result by substituting and combining the transformation equations, but we can also see the result directly from the principle of relativity for the following reason (it is always good to look again to see whether we can see the reason). We have already (Figure 3-3) discussed how a possible clock might work when it is moving; the light appears to travel at an angle at the speed c in the fixed system, while it simply goes vertically with the same speed in the moving system. We found that the *vertical component* of the velocity in the fixed system is less than that of light by the factor $\sqrt{1 - u^2/c^2}$ (see Eq. 3-3). But now suppose that we let a material particle go back and forth in this same "clock," but at some integral fraction $1/n$ of the speed of light (Figure 4-1). Then when the particle has gone back and forth once, the light will have gone exactly n times. That is, each "click" of the "particle" clock will coincide with each nth

Relativistic Energy and Momentum

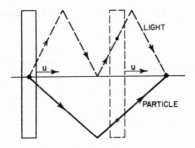

Figure 4-1. Trajectories described by a light ray and particle inside a moving clock.

"click" of the light clock. *This fact must still be true when the whole system is moving*, because the physical phenomenon of coincidence will be a coincidence in any frame. Therefore, since the speed c_y is less than the speed of light, the speed v_y of the particle must be slower than the corresponding speed by the same square-root ratio! That is why the square root appears in any vertical velocity.

4-4 Relativistic mass

We learned in the last chapter that the mass of an object increases with velocity, but no demonstration of this was given, in the sense that we made no arguments analogous to those about the way clocks have to behave. However, we *can* show that, as a consequence of relativity plus a few other reasonable assumptions, the mass must vary in this way. (We have to say "a few other assumptions" because we cannot prove anything unless we have some laws which we assume to be true, if we expect to make meaningful deductions.) To avoid the need to study the transformation laws of force, we shall analyze a *collision*, where we need know nothing about the laws of force, except that we shall assume the conservation of momentum and energy. Also, we shall assume that the momentum of a particle which is moving is a vector and is always directed in the direction of the velocity. However, we shall not assume that the momentum is a *constant* times the velocity, as Newton did, but only that it is some *function* of velocity. We thus write the momentum vector as a certain coefficient times the vector velocity:

$$\mathbf{p} = m_v \mathbf{v}. \qquad (4.8)$$

We put a subscript v on the coefficient to remind us that it is a function of velocity, and we shall agree to call this coefficient m_v the "mass." Of course, when the velocity is small, it is the same mass that we would measure in the slow-moving experiments that we are used to. Now we shall try to demonstrate that the formula for m_v must be $m_0/\sqrt{1 - v^2/c^2}$, by arguing from the principle of relativity that the laws of physics must be the same in every coordinate system.

Suppose that we have two particles, like two protons, that are absolutely equal, and they are moving toward each other with exactly equal velocities. Their total momentum is zero. Now what can happen? After the collision, their directions of motion must be exactly opposite to each other, because if they are not exactly opposite, there will be a nonzero total vector momentum, and momentum would not have been conserved. Also they must have the same speeds, since they are exactly similar objects; in fact, they must have the same speed they started with, since we suppose that the energy is conserved in these collisions. So the diagram of an elastic collision, a reversible collision, will look like Figure 4-2(a): all the arrows are the same length, all the speeds are equal. We shall suppose that such collisions can always be arranged, that any angle θ can occur, and that any speed could be used in such a collision. Next, we notice that this same collision can be viewed differently by turning the axes, and just for convenience we *shall* turn the axes, so that the horizontal splits it evenly, as in Figure 4-2(b). It is the same collision redrawn, only with the axes turned.

Now here is the real trick: let us look at this collision from the point of view of someone riding along in a car that is moving with a speed equal to the horizontal component of the velocity of one particle.

Relativistic Energy and Momentum

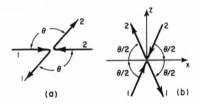

Figure 4-2. Two views of an elastic collision between equal objects moving at the same speed in opposite directions.

Then how does the collision look? It looks as though particle 1 is just going straight up, because it has lost its horizontal component, and it comes straight down again, also because it does not have that component. That is, the collision appears as shown in Fig. 4-3(a). Particle 2, however, was going the other way, and as we ride past it appears to fly by at some terrific speed and at a smaller angle, but we can appreciate that the angles before and after the collision are the *same*. Let us denote by u the horizontal component of the velocity of particle 2, and by w the vertical velocity of particle 1.

Now the question is, what is the vertical velocity $u \tan \alpha$? If we knew that, we could get the correct expression for the momentum, using the law of conservation of momentum in the vertical direction. Clearly, the horizontal component of the momentum is conserved: it is the same before and after the collision for both particles, and is zero for particle 1. So we need use the conservation law only for the upward velocity $u \tan \alpha$. *But we can* get the upward velocity, simply by looking at the same collision going the other way! If we look at the collision of Figure 4-3(a) from a car to the left moving with speed u, we see the same collision, except "turned over," as shown in Figure 4-3(b). Now particle 2 is the one that goes up and down with speed w, and particle 1 has picked up the horizontal speed u. Of course, now we *know* what the velocity $u \tan \alpha$ is: it is $w \sqrt{1 - u^2/c^2}$ (see Eq. 4.7). We know that the change in the vertical momentum of the vertically moving particle is

$$\Delta p = 2m_w w$$

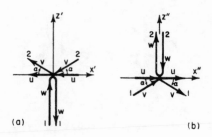

Figure 4-3. Two more views of the collision, from moving cars.

(2, because it moves up and back down). The obliquely moving particle has a certain velocity v whose components we have found to be u and $w\sqrt{1 - u^2/c^2}$, and whose mass is m_v. The change in *vertical* momentum of this particle is therefore $\Delta p' = 2m_v w \sqrt{1 - u^2/c^2}$ because, in accordance with our assumed law (4.8), the momentum component is always the mass corresponding to the magnitude of the velocity times the component of the velocity in the direction of interest. Thus in order for the total momentum to be zero the vertical momenta must cancel and the ratio of the mass moving with speed v and the mass moving with speed w must therefore be

$$\frac{m_w}{m_v} = \sqrt{1 - u^2/c^2}. \tag{4.9}$$

Let us take the limiting case that w is infinitesimal. If w is very tiny indeed, it is clear that v and u are practically equal. In this case, $m_w \to m_0$ and $m_v \to m_u$. The grand result is

$$m_u = \frac{m_0}{\sqrt{1 - u^2/c^2}}. \tag{4.10}$$

It is an interesting exercise now to check whether or not Eq. (4.9) is indeed true for arbitrary values of w, assuming that Eq. (4.10) is the right formula for the mass. Note that the velocity v needed in Eq. (4.9) can be calculated from the right-angle triangle:

$$v^2 = u^2 + w^2(1 - u^2/c^2).$$

Relativistic Energy and Momentum

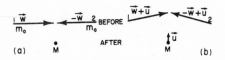

Figure 4-4. Two views of an inelastic collision between equally massive objects.

It will be found to check out automatically, although we used it only in the limit of small w.

Now, let us accept that momentum is conserved and that the mass depends upon the velocity according to (4.10) and go on to find what else we can conclude. Let us consider what is commonly called an *inelastic collision*. For simplicity, we shall suppose that two objects of the same kind, moving oppositely with equal speeds w, hit each other and stick together, to become some new, stationary object, as shown in Figure 4-4(a). The mass m of each corresponds to w, which, as we know, is $m_0/\sqrt{1 - w^2/c^2}$. If we assume the conservation of momentum and the principle of relativity, we can demonstrate an interesting fact about the mass of the new object which has been formed. We imagine an infinitesimal velocity u at right angles to w (we can do the same with finite values of u, but it is easier to understand with an infinitesimal velocity), then look at this same collision as we ride by in an elevator at the velocity $-u$. What we see is shown in Figure 4-4(b). The composite object has an unknown mass M. Now object 1 moves with an upward component of velocity u and a horizontal component which is practically equal to w, and so also does object 2. After the collision we have the mass M moving upward with velocity u, considered very small compared with the speed of light, and also small compared with w. Momentum must be conserved, so let us estimate the momentum in the upward direction before and after the collision. Before the collision we have $p \sim 2m_w u$, and after the collision the momentum is evidently $p' = M_u u$, but M_u is essentially the same as M_0 because u is so small. These momenta must be equal because of the conservation of momentum, and therefore

$$M_0 = 2m_w. \tag{4.11}$$

The mass of the object which is formed when two equal objects collide must be twice the mass of the objects which come together. You might say, "Yes, of course, that is the conservation of mass." But not "Yes, of course," so easily, because *these masses have been enhanced* over the masses that they would be if they were standing still, yet they still contribute, to the total *M*, not the mass they have when standing still, but *more*. Astonishing as that may seem, in order for the conservation of momentum to work when two objects come together, the mass that they form must be greater than the rest masses of the objects, even though the objects are at rest after the collision!

4-5 Relativistic energy

In the last chapter we demonstrated that as a result of the dependence of the mass on velocity and Newton's laws, the changes in the kinetic energy of an object resulting from the total work done by the forces on it always comes out to be

$$\Delta T = (m_u - m_0)c^2 = \frac{m_0 c^2}{\sqrt{1 - u^2/c^2}} - m_0 c^2. \qquad (4.12)$$

We even went further, and guessed that the total energy is the total mass times c^2. Now we continue this discussion.

Suppose that our two equally massive objects that collide can still be "seen" inside *M*. For instance, a proton and a neutron are "stuck together," but are still moving about inside of *M*. Then, although we might at first expect the mass *M* to be $2m_0$, we have found that it is not $2m_0$, but $2m_w$. Since $2m_w$ is what is put in, but $2m_0$ are the rest masses of the things inside, the *excess* mass of the composite object is equal to the kinetic energy brought in. This means, of course, that *energy has inertia*. In the last chapter we discussed the heating of a gas, and showed that because the gas molecules are moving and moving things are heavier, when we put energy into the gas its

Relativistic Energy and Momentum

molecules move faster and so the gas gets heavier. But in fact the argument is completely general, and our discussion of the inelastic collision shows that the mass is there whether or not it is *kinetic* energy. In other words, if two particles come together and produce potential or any other form of energy; if the pieces are slowed down by climbing hills, doing work against internal forces, or whatever; then it is still true that the mass is the total energy that has been put in. So we see that the conservation of mass which we have deduced above is equivalent to the conservation of energy, and therefore there is no place in the theory of relativity for strictly inelastic collisions, as there was in Newtonian mechanics. According to Newtonian mechanics it is all right for two things to collide and so form an object of mass $2m_0$ which is in no way distinct from the one that would result from putting them together slowly. Of course we know from the law of conservation of energy that there is more kinetic energy inside, but that does not affect the mass, according to Newton's laws. But now we see that this is impossible; because of the kinetic energy involved in the collision, the resulting object will be *heavier*; therefore, it will be a *different* object. When we put the objects together gently they make something whose mass is $2m_0$; when we put them together forcefully, they make something whose mass is greater. When the mass is different, we can *tell* that it is different. So, necessarily, the conservation of energy must go along with the conservation of momentum in the theory of relativity.

This has interesting consequences. For example, suppose that we have an object whose mass M is measured, and suppose something happens so that it flies into two equal pieces moving with speed w, so that they each have a mass m_w. Now suppose that these pieces encounter enough material to slow them up until they stop; then they will have mass m_0. How much energy will they have given to the material when they have stopped? Each will give an amount $(m_w - m_0)c^2$, by the theorem that we proved before. This much energy is left in the material in some form, as heat, potential energy, or whatever. Now $2m_w = M$, so the liberated energy is $E = (M - 2m_0)c^2$. This equation was used to estimate how much energy would be liberated under fission in the atomic bomb, for example. (Although the fragments are not exactly equal, they are nearly equal.) The mass of the uranium atom was known—it had been measured ahead of time—and the atoms into which it split, iodine, xenon, and so on, all were of known mass. By masses, we do not mean the masses while the atoms are moving, we mean the masses when the atoms are *at rest*. In other words, both M and m_0 are known. So by subtracting the two numbers one can calculate how much energy will be released if M can be made to split in "half." For this reason poor old Einstein was called the "father" of the atomic bomb in all the newspapers. Of course, all that meant was that he could tell us ahead of time how much energy would be released if we told him what process would occur. The energy that should be liberated when an atom of uranium undergoes fission was estimated about six months before the first direct test, and as soon as the energy was in fact liberated, someone measured it directly (and if Einstein's formula had not worked, they would have measured it anyway), and the moment they measured it they no longer needed the formula. Of course, we should not belittle Einstein, but rather should criticize the newspapers and many popular descriptions of what causes what in the history of physics and technology. The problem of how to get the thing to occur in an effective and rapid manner is a completely different matter.

The result is just as significant in chemistry. For instance, if we were to weigh the carbon dioxide molecule and compare its mass with that of the carbon and the oxygen, we could find out how much energy would be liberated when carbon and oxygen form carbon dioxide. The only trouble here is that the differences in masses are so small that it is technically very difficult to do.

Now let us turn to the question of whether we should add m_0c^2 to the kinetic energy and say from now on that the total energy of an object is mc^2. First, if we can still *see* the component pieces of rest mass m_0 inside M, then we could say that some of the mass M of the compound object is the mechanical rest mass of the parts, part of it is kinetic energy of the parts, and part of it is potential energy of the

Relativistic Energy and Momentum

parts. But we have discovered, in nature, particles of various kinds which undergo reactions just like the one we have treated above, in which with all the study in the world, *we cannot see the parts inside*. For instance, when a K-meson disintegrates into two pions it does so according to the law (4.11), but the idea that a K is made out of 2 π's is a useless idea, because it also disintegrates into 3 π's!

Therefore we have a *new idea*: we do not have to know what things are made of inside; we cannot and need not identify, inside a particle, which of the energy is rest energy of the parts into which it is going to disintegrate. It is not convenient and often not possible to separate the total mc^2 energy of an object into rest energy of the inside pieces, kinetic energy of the pieces, and potential energy of the pieces; instead, we simply speak of the *total energy* of the particle. We "shift the origin" of energy by adding a constant $m_0 c^2$ to everything, and say that the total energy of a particle is the mass in motion times c^2, and when the object is standing still, the energy is the mass at rest times c^2.

Finally, we find that the velocity v, momentum P, and total energy E are related in a rather simple way. That the mass in motion at speed v is the mass m_0 at rest divided by $\sqrt{1 - v^2/c^2}$, surprisingly enough, is rarely used. Instead, the following relations are easily proved, and turn out to be very useful:

$$E^2 - P^2 c^2 = m_0^2 c^4 \tag{4.13}$$

and

$$Pc = Ev/c. \tag{4.14}$$

Five

SPACE-TIME

5-1 The geometry of space-time

The theory of relativity shows us that the relationships of positions and times as measured in one coordinate system and another are not what we would have expected on the basis of our intuitive ideas. It is very important that we thoroughly understand the relations of space and time implied by the Lorentz transformation, and therefore we shall consider this matter more deeply in this chapter.

The Lorentz transformation between the positions and times (x, y, z, t) as measured by an observer "standing still," and the corresponding coordinates and time (x', y', z', t') measured inside a "moving" space ship, moving with velocity u are

$$x' = \frac{x - ut}{\sqrt{1 - u^2/c^2}},$$

$$y' = y,$$

$$z' = z,$$

$$t' = \frac{t - ux/c^2}{\sqrt{1 - u^2/c^2}}.$$

(5.1)

Let us compare these equations with Eq. (1.5), which also relates measurements in two systems, one of which in this instance is *rotated* relative to the other:

SIX NOT-SO-EASY PIECES

$$x' = x \cos \theta + y \sin \theta,$$
$$y' = y \cos \theta - x \sin \theta, \qquad (5.2)$$
$$z' = z.$$

In this particular case, Moe and Joe are measuring with axes having an angle θ between the x'- and x-axes. In each case, we note that the "primed" quantities are "mixtures" of the "unprimed" ones: the new x' is a mixture of x and y, and the new y' is also a mixture of x and y.

An analogy is useful: When we look at an object, there is an obvious thing we might call the "apparent width," and another we might call the "depth." But the two ideas, width and depth, are not *fundamental* properties of the object, because if we step aside and look at the same thing from a different angle, we get a different width and a different depth, and we may develop some formulas for computing the new ones from the old ones and the angles involved. Equations (5.2) are these formulas. One might say that a given depth is a kind of "mixture" of all depth and all width. If it were impossible ever to move, and we always saw a given object from the same position, then this whole business would be irrelevant—we would always see the "true" width and the "true" depth, and they would appear to have quite different qualities, because one appears as a subtended optical angle and the other involves some focusing of the eyes or even intuition; they would seem to be very different things and would never get mixed up. It is because we *can* walk around that we realize that depth and width are, somehow or other, just two different aspects of the same thing.

Can we not look at the Lorentz transformations in the same way? Here also we have a mixture—of positions and the time. A difference between a space measurement and a time measurement produces a new space measurement. In other words, in the space measurements of one man there is mixed in a little bit of the time, as seen by the other. Our analogy permits us to generate this idea: The "reality" of an object that we are looking at is somehow greater (speaking crudely and intuitively) than its "width" and its "depth" because *they* depend upon *how* we look at it; when we move to a new position, our brain immediately recalculates the width and the depth. But our brain does not immediately recalculate coordinates and time when we move at high speed, because we have had no effective experience of going nearly as fast as light to appreciate the fact that time and space are also of the same nature. It is as though we were always stuck in the position of having to look at just the width of something, not being able to move our heads appreciably one way or the other; if we could, we understand now, we would see some of the other man's time—we would see "behind," so to speak, a little bit.

Thus we shall try to think of objects in a new kind of world, of space and time mixed together, in the same sense that the objects in our ordinary space-world are real, and can be looked at from different directions. We shall then consider that objects occupying space and lasting for a certain length of time occupy a kind of a "blob" in a new kind of world, and that we look at this "blob" from different points of view when we are moving at different velocities. This new world, this geometrical entity in which the "blobs" exist by occupying position and taking up a certain amount of time, is called *space-time*. A given point (x, y, z, t) in space-time is called an *event*. Imagine, for example, that we plot the x-positions horizontally, y and z in two other directions, both mutually at "right angles" and at "right angles" to the paper (!), and time, vertically. Now, how does a moving particle, say, look on such a diagram? If the particle is standing still, then it has a certain x, and as time goes on, it has the same x, the same x, the same x; so its "path" is a line that runs parallel to the t-axis (Figure 5-1 (a)). On the other hand, if it drifts outward, then as the time goes on x increases (Figure 5-1 (b)). So a particle, for example, which starts to drift out and then slows up should have a motion something like that shown in Figure 5-1 (c). A particle, in other words, which is permanent and does not disintegrate is represented by a line in space-time. A particle which disintegrates would be represented by a forked line, because it would turn into two other things which would start from that point.

SIX NOT-SO-EASY PIECES

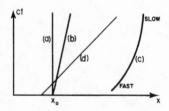

Figure 5-1. Three particle paths in space-time: (a) a particle at rest at
$x = x_0$; (b) a particle which starts at $x = x_0$ and moves
with constant speed; (c) a particle which starts at high
speed but slows down.

What about light? Light travels at the speed c, and that would be
represented by a line having a certain fixed slope (Figure 5-1 (d)).
Now according to our new idea, if a given event occurs to a
particle, say if it suddenly disintegrates at a certain space-time point
into two new ones which follow some new tracks, and this interesting event occurred at a certain value of x and a certain value of t,
then we would expect that, if this makes any sense, we just have to
take a new pair of axes and turn them, and that will give us the new t
and the new x in our new system, as shown in Figure 5-2(a). But this
is wrong, because Eq. (5.1) is not *exactly* the same mathematical
transformation as Eq. (5.2). Note, for example, the difference in
sign between the two, and the fact that one is written in terms of cos
θ and sin θ, while the other is written with algebraic quantities. (Of
course, it is not impossible that the algebraic quantities could be
written as cosine and sine, but actually they cannot.) But still, the

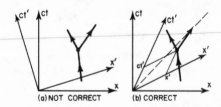

Figure 5-2. Two views of a disintegrating particle.

Space-Time

two expressions *are* very similar. As we shall see, it is not really possible to think of space-time as a real, ordinary geometry because of that difference in sign. In fact, although we shall not emphasize this point, it turns out that a man who is moving has to use a set of axes which are inclined equally to the light ray, using a special kind of projection parallel to the x'- and t'-axes, for his x' and t', as shown in Figure 5-2(b). We shall not deal with the geometry, since it does not help much; it is easier to work with the equations.

5-2 Space-time intervals

Although the geometry of space-time is not Euclidean in the ordinary sense, there *is* a geometry which is very similar, but peculiar in certain respects. If this idea of geometry is right, there ought to be some functions of coordinates and time which are independent of the coordinate system. For example, under ordinary rotations, if we take two points, one at the origin, for simplicity, and the other one somewhere else, both systems would have the same origin, and the distance from here to the other point is the same in both. That is one property that is independent of the particular way of measuring it. The square of the distance is $x^2 + y^2 + z^2$. Now what about space-time? It is not hard to demonstrate that we have here, also, something which stays the same, namely, the combination $c^2t^2 - x^2 - y^2 - z^2$ is the same before and after the transformation:

$$c^2t'^2 - x'^2 - y'^2 - z'^2 = c^2t^2 - x^2 - y^2 - z^2. \tag{5.3}$$

This quantity is therefore something which, like the distance, is "real" in some sense; it is called the *interval* between the two space-time points, one of which is, in this case, at the origin. (Actually, of course, it is the interval squared, just as $x^2 + y^2 + z^2$ is the distance squared.) We give it a different name because it is in a different geometry, but the interesting thing is only that some signs are reversed and there is a c in it.

Let us get rid of the c; that is an absurdity if we are going to have a wonderful space with x's and y's that can be interchanged. One of the confusions that could be caused by someone with no experience would be to measure widths, say, by the angle subtended at the eye, and measure depth in a different way, like the strain on the muscles needed to focus them, so that the depths would be measured in feet and the widths in meters. Then one would get an enormously complicated mess of equations in making transformations such as (5.2), and would not be able to see the clarity and simplicity of the thing for a very simple technical reason, that the same thing is being measured in two different units. Now in Eqs. (5.1) and (5.3) nature is telling us that time and space are equivalent; time becomes space; *they should be measured in the same units*. What distance is a "second"? It is easy to figure out from (5.3) what it is. It is 3×10^8 meters, *the distance that light would go in one second*. In other words, if we were to measure all distances and times in the same units, seconds, then our unit of distance would be 3×10^8 meters, and the equations would be simpler. Or another way that we could make the units equal is to measure time in meters. What is a meter of time? A meter of time is the time it takes for light to go one meter, and is therefore $\frac{1}{3} \times 10^{-8}$ sec, or 3.3 billionths of a second! We would like, in other words, to put all our equations in a system of units in which $c = 1$. If time and space are measured in the same units, as suggested, then the equations are obviously much simplified. They are

$$x' = \frac{x - ut}{\sqrt{1 - u^2}} \, ,$$
$$y' = y,$$
$$z' = z, \tag{5.4}$$
$$t' = \frac{t - ux}{\sqrt{1 - u^2}} \, ,$$

$$t'^2 - x'^2 - y'^2 - z'^2 = t^2 - x^2 - y^2 - z^2. \tag{5.5}$$

If we are ever unsure or "frightened" that after we have this system with $c = 1$ we shall never be able to get our equations right again, the answer is quite the opposite. It is much easier to remember them without the c's in them, and it is always easy to put the c's back, by looking after the dimensions. For instance, in $\sqrt{1 - u^2}$, we know that we cannot subtract a velocity squared, which has units, from the pure number 1, so we know that we must divide u^2 by c^2 in order to make that unitless, and that is the way it goes.

The difference between space-time and ordinary space, and the character of an interval as related to the distance, is very interesting. According to formula (5.5), if we consider a point which in a given coordinate system had zero time, and only space, then the interval squared would be negative and we would have an imaginary interval, the square root of a negative number. Intervals can be either real or imaginary in the theory. The square of an interval may be either positive or negative, unlike distance, which has a positive square. When an interval is imaginary, we say that the two points have a *space-like interval* between them (instead of imaginary), because the interval is more like space than like time. On the other hand, if two objects are at the same place in a given coordinate system, but differ only in time, then the square of the time is positive and the distances are zero and the interval squared is positive; this is called a *time-like interval*. In our diagram of space-time, therefore, we would have a representation something like this: at 45° there are two lines (actually, in four dimensions these will be "cones," called light cones) and points on these lines are all at zero interval from the origin. Where light goes from a given point is always separated from it by a zero interval, as we see from Eq. (5.5). Incidentally, we have just proved that if light travels with speed c in one system, it travels with speed c in another, for if the interval is the same in both systems, i.e., zero in one and zero in the other, then to state that the propagation speed of light is invariant is the same as saying that the interval is zero.

5-3 Past, present, and future

The space-time region surrounding a given space-time point can be separated into three regions, as shown in Figure 5-3. In one region we have space-like intervals, and in two regions, time-like intervals.

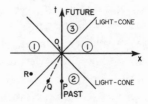

Figure 5-3. The space-time region surrounding a point at the origin.

Physically, these three regions into which space-time around a given point is divided have an interesting physical relationship to that point: a physical object or a signal can get from a point in region 2 to the event O by moving along at a speed less than the speed of light. Therefore events in this region can affect the point O, can have an influence on it from the past. In fact, of course, an object at P on the negative t-axis is precisely in the "past" with respect to O; it is the same space-point as O, only earlier. What happened there then, affects O now. (Unfortunately, that is the way life is.) Another object at Q can get to O by moving with a certain speed less than c, so if this object were in a space ship and moving, it would be, again, the past of the same space-point. That is, in another coordinate system, the axis of time might go through both O and Q. So all points of region 2 are in the "past" of O, and anything that happens in this region *can* affect O. Therefore region 2 is sometimes called the *affective past*, or affecting past; it is the locus of all events which can affect point O in any way.

Region 3, on the other hand, is a region which we can affect *from* O, we can "hit" things by shooting "bullets" out at speeds less than c. So this is the world whose future can be affected by us, and we may call that the *affective future*. Now the interesting thing about all the rest of space-time, i.e., region 1, is that we can neither affect it now *from* O, nor can it affect us now *at* O, because nothing can go faster than the speed of light. Of course, what happens at R *can* affect us *later* that is, if the sun is exploding "right now," it takes eight minutes before we know about it, and it cannot possibly affect us before then.

What we mean by "right now" is a mysterious thing which we cannot define and we cannot affect, but it can affect us later, or we could have affected it if we had done something far enough in the past. When we look at the star Alpha Centauri, we see it as it was four years ago; we might wonder what it is like "now." "Now" means at the same time from our special coordinate system. We can only see Alpha Centauri by the light that has come from our past, up to four years ago, but we do not know what it is doing "now"; it will take four years before what it is doing "now" can affect us. Alpha Centauri "now" is an idea or concept of our mind; it is not something that is really definable physically at the moment, because we have to wait to observe it; we cannot even define it right "now." Furthermore, the "now" depends on the coordinate system. If, for example, Alpha Centauri were moving, an observer there would not agree with us because he would put his axes at an angle, and his "now" would be a *different* time. We have already talked about the fact that simultaneity is not a unique thing.

There are fortune tellers, or people who tell us they can know the future, and there are many wonderful stories about the man who suddenly discovers that he has knowledge about the affective future. Well, there are lots of paradoxes produced by that because if we know something is going to happen, then we can make sure we will avoid it by doing the right thing at the right time, and so on. But actually there is no fortune teller who can even tell us the *present*! There is no one who can tell us what is really happening right now, at any reasonable distance, because that is unobservable. We might ask ourselves this question, which we leave to the student to try to answer: Would any paradox be produced if it were suddenly to become possible to know things that are in the space-like intervals of region 1?

5-4 More about four-vectors

Let us now return to our consideration of the analogy of the Lorentz transformation and rotations of the space axes. We have learned the utility of collecting together other quantities which have the same transformation properties as the coordinates, to form what we call *vectors*, directed lines. In the case of ordinary rotations, there are many quantities that transform the same way as x, y, and z under rotation: for example, the velocity has three components, an x, y, and z component; when seen in a different coordinate system, none of the components is the same, instead they are all transformed to new values. But, somehow or other, the velocity "itself" has a greater reality than do any of its particular components, and we represent it by a directed line.

We therefore ask: Is it or is it not true that there are quantities which transform, or which are related, in a moving system and in a nonmoving system, in the same way as x, y, z, and t? From our experience with vectors, we know that three of the quantities, like x, y, z, would constitute the three components of an ordinary space-vector, but the fourth quantity would look like an ordinary scalar under space rotation, because it does not change so long as we do not go into a moving coordinate system. Is it possible, then, to associate with some of our known "three-vectors" a fourth object, that we could call the "time component," in such a manner that the four objects together would "rotate" the same way as position and time in space-time? We shall now show that there is, indeed, at least one such thing (there are many of them, in fact): *the three components of momentum, and the energy as the time component, transform together* to make what we call a "four-vector." In demonstrating this, since it is quite inconvenient to have to write c's everywhere, we shall use the same trick concerning units of the energy, the mass, and the momentum, that we used in Eq. (5.4). Energy and mass, for example, differ only by a factor c^2 which is merely a question of units, so we can say energy *is* the mass. Instead

Space-Time

of having to write the c^2, we put $E = m$, and then, of course, if there were any trouble we would put in the right amounts of c so that the units would straighten out in the last equation, but not in the intermediate ones.

Thus our equations for energy and momentum are

$$E = m = m_0/\sqrt{1 - v^2},$$
$$\mathbf{p} = m\mathbf{v} = m_0\mathbf{v}/\sqrt{1 - v^2}. \tag{5.6}$$

Also in these units, we have

$$E^2 - p^2 = m_0^2. \tag{5.7}$$

For example, if we measure energy in electron volts, what does a mass of 1 electron volt mean? It means the mass whose rest energy is 1 electron volt, that is, m_0c^2 is one electron volt. For example, the rest mass of an electron is 0.511×10^6 ev.

Now what would the momentum and energy look like in a new coordinate system? To find out, we shall have to transform Eq. (5.6), which we can do because we know how the velocity transforms. Suppose that, as we measure it, an object has a velocity v, but we look upon the same object from the point of view of a space ship which itself is moving with a velocity u, and in that system we use a prime to designate the corresponding thing. In order to simplify things at first, we shall take the case that the velocity v is in the direction of u. (Later, we can do the more general case.) What is v', the velocity as seen from the space ship? It is the composite velocity, the "difference" between v and u. By the law which we worked out before,

$$v' = \frac{v - u}{1 - uv}. \tag{5.8}$$

Now let us calculate the new energy E', the energy as the fellow in the space ship would see it. He would use the same rest mass, of course, but he would use v' for the velocity. What we have to do is square v', subtract it from one, take the square root, and take the reciprocal:

$$v'^2 = \frac{v^2 - 2uv + u^2}{1 - 2uv + u^2v^2},$$

$$1 - v'^2 = \frac{1 - 2uv + u^2v^2 - v^2 + 2uv - u^2}{1 - 2uv + u^2v^2}$$

$$= \frac{1 - v^2 - u^2 + u^2v^2}{1 - 2uv + u^2v^2}$$

$$= \frac{(1 - v^2)(1 - u^2)}{(1 - uv)^2}.$$

Therefore

$$\frac{1}{\sqrt{1 - v'^2}} = \frac{1 - uv}{\sqrt{1 - v^2}\,\sqrt{1 - u^2}}. \tag{5.9}$$

The energy E' is then simply m_0 times the above expression. But we want to express the energy in terms of the unprimed energy and momentum, and we note that

$$E' = \frac{m_0 - m_0 uv}{\sqrt{1 - v^2}\,\sqrt{1 - u^2}} = \frac{(m_0/\sqrt{1 - v^2}) - (m_0 v/\sqrt{1 - v^2})u}{\sqrt{1 - u^2}},$$

or

$$E' = \frac{E - up_x}{\sqrt{1 - u^2}}, \tag{5.10}$$

which we recognize as being exactly of the same form as

$$t' = \frac{t - ux}{\sqrt{1 - u^2}}.$$

Next we must find the new momentum p'_x. This is just the energy E times v', and is also simply expressed in terms of E and p:

$$p'_x = E'v' = \frac{m_0(1 - uv)}{\sqrt{1 - v^2}\,\sqrt{1 - u^2}} \cdot \frac{v - u}{(1 - uv)} = \frac{m_0 v - m_0 u}{\sqrt{1 - v^2}\,\sqrt{1 - u^2}}.$$

Thus

$$p'_x = \frac{p_x - uE}{\sqrt{1 - u^2}}, \tag{5.11}$$

Space-Time

which we recognize as being of precisely the same form as

$$x' = \frac{x - ut}{\sqrt{1 - u^2}} \cdot$$

Thus the transformations for the new energy and momentum in terms of the old energy and momentum are exactly the same as the transformations for t' in terms of t and x, and x' in terms of x and t: all we have to do is, every time we see t in (5.4) substitute E, and every time we see x substitute p_x, and then the equations (5.4) will become the same as Eqs. (5.10) and (5.11). This would imply, if everything works right, an additional rule that $p'_y = p_y$ and that $p'_z = p_z$. To prove this would require our going back and studying the case of motion up and down. Actually, we did study the case of motion up and down in the last chapter. We analyzed a complicated collision and we noticed that, in fact, the transverse momentum is *not* changed when viewed from a moving system; so we have already verified that $p'_y = p_y$ and $p'_z = p_z$. The complete transformation, then, is

$$
\begin{aligned}
p'_x &= \frac{p_x - uE}{\sqrt{1 - u^2}}, \\
p'_y &= p_y, \\
p'_z &= p_z, \\
E' &= \frac{E - up_x}{\sqrt{1 - u^2}} \cdot
\end{aligned}
\tag{5.12}
$$

In these transformations, therefore, we have discovered four quantities which transform like x, y, z, and t, and which we call the *four-vector momentum*. Since the momentum is a four-vector, it can be represented on a space-time diagram of a moving particle as an "arrow" tangent to the path, as shown in Figure 5-4. This arrow has a time component equal to the energy, and its space components represent its three-vector momentum; this arrow is more "real" than either the energy or the momentum, because those just depend on how we look at the diagram.

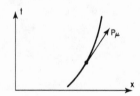

Figure 5-4. The four-vector momentum of a particle.

5-5 Four-vector algebra

The notation for four-vectors is different than it is for three-vectors. In the case of three-vectors, if we were to talk about the ordinary three-vector momentum we would write it **p**. If we wanted to be more specific, we could say it has three components which are, for the axes in question, p_x, p_y, and p_z, or we could simply refer to a general component as p_i, and say that i could either be x, y, or z, and that these are the three components; that is, imagine that i is any one of three directions, x, y, or z. The notation that we use for four-vectors is analogous to this: we write p_μ for the four-vector, and μ stands for the *four* possible directions t, x, y, or z.

We could, of course, use any notation we want; do not laugh at notations; invent them, they are powerful. In fact, mathematics is, to a large extent, invention of better notations. The whole idea of a four-vector, in fact, is an improvement in notation so that the transformations can be remembered easily. A_μ, then, is a general four-vector, but for the special case of momentum, the p_t is identified as the energy, p_x is the momentum in the x-direction, p_y is that in the y-direction, and p_z is that in the z-direction. To add four-vectors, we add the corresponding components.

If there is an equation among four-vectors, then the equation is true for *each component*. For instance, if the law of conservation of three-vector momentum is to be true in particle collisions, i.e., if the sum of the momenta for a large number of interacting or colliding particles is to be a constant, that must mean that the sums of all momenta in the x-direction, in the y-direction, and in the z-

Space-Time

direction, for all the particles, must each be constant. This law alone would be impossible in relativity because it is *incomplete*; it is like talking about only two of the components of a three-vector. It is incomplete because if we rotate the axes, we mix the various components, so we must include all three components in our law. Thus, in relativity, we must complete the law of conservation of momentum by extending it to include the *time* component. This is *absolutely necessary* to go with the other three, or there cannot be relativistic invariance. The *conservation of energy* is the fourth equation which goes with the conservation of momentum to make a valid four-vector relationship in the geometry of space and time. Thus the law of conservation of energy and momentum in four-dimensional notation is

$$\sum_{\substack{particles \\ in}} p_\mu = \sum_{\substack{particles \\ out}} p_\mu, \tag{5.13}$$

or, in a slightly different notation

$$\sum_i p_{i\mu} = \sum_j p_{j\mu}, \tag{5.14}$$

where $i = 1, 2, \ldots$ refers to the particles going into the collision, $j = 1, 2, \ldots$ refers to the particles coming out of the collision, and $\mu = x, y, z$, or t. You say, "In which axes?" It makes no difference. The law is true for each component, using *any* axes.

In vector analysis we discussed one other thing, the dot product of two vectors. Let us now consider the corresponding thing in space-time. In ordinary rotation we discovered there was an unchanged quantity $x^2 + y^2 + z^2$. In four dimensions, we find that the corresponding quantity is $t^2 - x^2 - y^2 - z^2$ (Eq. 5.3). How can we write that? One way would be to write some kind of four-dimensional thing with a square dot between, like $A_\mu \otimes B_\mu$; one of the notations which is actually used is

$$\sideset{}{'}\sum_\mu A_\mu A_\mu = A_t^2 - A_x^2 - A_y^2 - A_z^2. \tag{5.15}$$

The prime on Σ means that the first term, the "time" term, is positive, but the other three terms have minus signs. This quantity, then, will be the same in any coordinate system, and we may call it the square of the length of the four-vector. For instance, what is the square of the length of the four-vector momentum of a single particle? This will be equal to $p_t^2 - p_x^2 - p_y^2 - p_z^2$ or, in other words, $E^2 - p^2$, because we know that p_t is E. What is $E^2 - p^2$? It must be something which is the same in every coordinate system. In particular, it must be the same for a coordinate system which is moving right along with the particle, in which the particle is standing still. If the particle is standing still, it would have no momentum. So in that coordinate system, it is purely its energy, which is the same as its rest mass. Thus $E^2 - p^2 = m_0^2$. So we see that the square of the length of this vector, the four-vector momentum, is equal to m_0^2.

From the square of a vector, we can go on to invent the "dot product," or the product which is a scalar: if a_μ is one four-vector and b_μ is another four-vector, then the scalar product is

$$\sum{}' a_\mu b_\mu = a_t b_t - a_x b_x - a_y b_y - a_z b_z. \tag{5.16}$$

It is the same in all coordinate systems.

Finally, we shall mention certain things whose rest mass m_0 is zero. A photon of light, for example. A photon is like a particle, in that it carries an energy and a momentum. The energy of a photon is a certain constant, called Planck's constant, times the frequency of the photon: $E = h\nu$. Such a photon also carries a momentum, and the momentum of a photon (or of any other particle, in fact) is h divided by the wavelength: $p = h/\lambda$. But, for a photon, there is a definite relationship between the frequency and the wavelength: $\nu = c/\lambda$. (The number of waves per second, times the wavelength of each, is the distance that the light goes in one second, which, of course, is c.) Thus we see immediately that the energy of a photon must be the momentum times c, or if $c = 1$, *the energy and momentum are equal*. That is to say, the rest mass is zero. Let us look at that

Space-Time

again; that is quite curious. If it is a particle of zero rest mass, what happens when it stops? *It never stops!* It always goes at the speed c. The usual formula for energy is $m_0/\sqrt{1 - v^2}$. Now can we say that $m_0 = 0$ and $v = 1$, so the energy is 0? We *cannot* say that it is zero; the photon really can (and does) have energy even though it has no rest mass, but this it possesses by perpetually going at the speed of light!

We also know that the momentum of any particle is equal to its total energy times its velocity: if $c = 1$, $p = vE$ or, in ordinary units, $p = vE/c^2$. For any particle moving at the speed of light, $p = E$ if $c = 1$. The formulas for the energy of a photon as seen from a moving system are, of course, given by Eq. (5.12), but for the momentum we must substitute the energy times c (or times 1 in this case). The different energies after transformation means that there are different frequencies. This is called the Doppler effect, and one can calculate it easily from Eq. (5.12), using also $E = p$ and $E = h\nu$.

As Minkowski said, "Space of itself, and time of itself will sink into mere shadows, and only a kind of union between them shall survive."

Six

CURVED SPACE

6-1 *Curved spaces with two dimensions*

◊ According to Newton, everything attracts everything else with a force inversely proportional to the square of the distance from it, and objects respond to forces with accelerations proportional to the forces. They are Newton's laws of universal gravitation and of motion. As you know, they account for the motions of balls, planets, satellites, galaxies, and so forth.

Einstein had a different interpretation of the law of gravitation. According to him, space and time—which must be put together as space-time—are *curved* near heavy masses. And it is the attempt of things to go along "straight lines" in this curved space-time which makes them move the way they do. Now that is a complex idea—very complex. It is the idea we want to explain in this chapter.

Our subject has three parts. One involves the effects of gravitation. Another involves the ideas of space-time which we already studied. The third involves the idea of curved space-time. We will simplify our subject in the beginning by not worrying about gravity and by leaving out the time—discussing just curved space. We will talk later about the other parts, but we will concentrate now on the idea of curved space—what is meant by curved space, and, more specifically, what is meant by curved space in this application of Einstein. Now even that much turns out to be somewhat difficult in three dimensions. So we will first reduce the problem still further and talk about what is meant by the words "curved space" in two dimensions.

In order to understand this idea of curved space in two dimensions you really have to appreciate the limited point of view of the character who lives in such a space. Suppose we imagine a bug with no eyes who lives on a plane, as shown in Figure 6-1. He can move only on the plane, and he has no way of knowing that there is any way to discover any "outside world." (He hasn't got your imagination.) We are, of course, going to argue by analogy. *We* live in a three-dimensional world, and we don't have any imagination about going off our three-dimensional world in a new direction; so we have to think the thing out by analogy. It is as though we were bugs living on a plane, and there was a space in another direction. That's why we will first work with the bug, remembering that he must live on his surface and can't get out.

As another example of a bug living in two dimensions, let's imagine one who lives on a sphere. We imagine that he can walk around on the surface of the sphere, as in Figure 6-2, but that he can't look "up," or "down," or "out."

Now we want to consider still a *third* kind of creature. He is also a bug like the others, and also lives on a plane, as our first bug did, but this time the plane is peculiar. The temperature is different at different places. Also, the bug and any rulers he uses are all made of

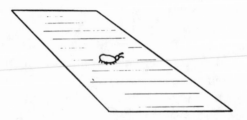

Figure 6-1. A bug on a plane surface.

Curved Space

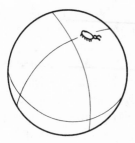

Figure 6-2. A bug on a sphere.

the same material which expands when it is heated. Whenever he puts a ruler somewhere to measure something the ruler expands immediately to the proper length for the temperature at that place. Wherever he puts any object—himself, a ruler, a triangle, or anything—the thing stretches itself because of the thermal expansion. Everything is longer in the hot places than it is in the cold places, and everything has the same coefficient of expansion. We will call the home of our third bug a "hot plate," although we will particularly want to think of a special kind of hot plate that is cold in the center and gets hotter as we go out toward the edges (Figure 6-3).

Now we are going to imagine that our bugs begin to study geometry. Although we imagine that they are blind so that they can't see any "outside" world, they can do a lot with their legs and feelers.

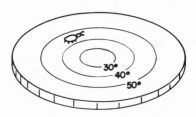

Figure 6-3. A bug on a hot plate.

SIX NOT-SO-EASY PIECES

They can draw lines, and they can make rulers, and measure off lengths. First, let's suppose that they start with the simplest idea in geometry. They learn how to make a straight line—defined as the shortest line between two points. Our first bug—see Figure 6-4—learns to make very good lines. But what happens to the bug on the sphere? He draws his straight line as the shortest distance—*for him*—between two points, as in Figure 6-5. It may look like a curve to us, but he has no way of getting off the sphere and finding out that there is "really" a shorter line. He just knows that if he tries any other path *in his world* it is always longer than his straight line. So we will let him have his straight line as the shortest arc between two points. (It is, of course an arc of a great circle.)

Finally, our third bug—the one in Figure 6-3—will also draw "straight lines" that look like curves to us. For instance, the shortest distance between *A* and *B* in Figure 6-6 would be on a curve like the one shown. Why? Because when his line curves out toward the warmer parts of his hot plate, the rulers get longer (from our omniscient point of view) and it takes fewer "yardsticks" laid end-to-end to get from *A* to *B*. So *for him* the line is straight—he has no way of knowing that there could be someone out in a strange three-dimensional world who would call a different line "straight."

We think you get the idea now that all the rest of the analysis will always be from the point of view of the creatures on the particular surfaces and not from *our* point of view. With that in mind let's see what the rest of their geometries looks like. Let's assume that the

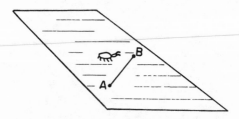

Figure 6-4. Making a "straight line" on a plane.

Curved Space

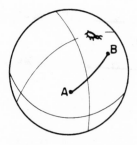

Figure 6-5. Making a "straight line" on a sphere.

bugs have all learned how to make two lines intersect at right angles. (You can figure out how they could do it.) Then our first bug (the one on the normal plane) finds an interesting fact. If he starts at the point *A* and makes a line 100 inches long, then makes a right angle and marks off another 100 inches, then makes another right angle and goes another 100 inches, then makes a third right angle and a fourth line 100 inches long, he ends up right at the starting point as shown in Figure 6-7(a). It is a property of his world—one of the facts of his "geometry."

Then he discovers another interesting thing. If he makes a triangle—a figure with three straight lines—the sum of the angles is equal to 180°, that is, to the sum of two right angles. See Figure 6-7(b).

Then he invents the circle. What's a circle? A circle is made this way: You rush off on straight lines in many many directions from a

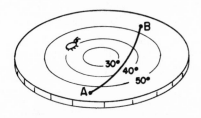

Figure 6-6. Making a "straight line" on the hot plate.

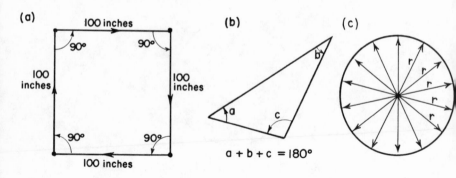

Figure 6-7. A square, triangle, and circle in a flat space.

single point, and lay out a lot of dots that are all the same distance from that point. See Figure 6-7(c). (We have to be careful how we define these things because we've got to be able to make the analogs for the other fellows.) Of course, it's equivalent to the curve you can make by swinging a ruler around a point. Anyway, our bug learns how to make circles. Then one day he thinks of measuring the distance around a circle. He measures several circles and finds a neat relationship: The distance around is always the same number times the radius r (which is, of course, the distance from the center out to the curve). The circumference and the radius always have the same ratio—approximately 6.283—independent of the size of the circle.

Now let's see what our other bugs have been finding out about *their* geometries. First, what happens to the bug on the sphere when he tries to make a "square"? If he follows the prescription we gave above, he would probably think that the result was hardly worth the trouble. He gets a figure like the one shown in Figure 6-8. His endpoint *B* isn't on top of the starting point *A*. It doesn't work out to a closed figure at all. Get a sphere and try it. A similar thing would happen to our friend on the hot plate. If he lays out four straight lines of equal length—as measured with his expanding rulers—joined by right angles he gets a picture like the one in Figure 6-9.

Curved Space

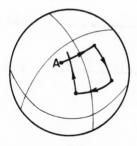

Figure 6-8. Trying to make a "square" on a sphere.

Now suppose that our bugs had each had their own Euclid who had told them what geometry "should" be like, and that they had checked him out roughly by making crude measurements on a *small* scale. Then as they tried to make accurate squares on a larger scale they would discover that something was wrong. The point is, that just by *geometrical measurements* they would discover that something was the matter with their space. We define a *curved space* to be a space in which the geometry is not what we expect for a plane. The geometry of the bugs on the sphere or on the hot plate is the geometry of a curved space. The rules of Euclidian geometry fail. And it isn't necessary to be able to lift yourself out of the plane in order to find out that the world that you live in is curved. It isn't necessary to circumnavigate the globe in order to find out that it is a ball. You can find out that you live

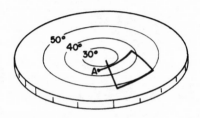

Figure 6-9. Trying to make a "square" on the hot plate.

on a ball by laying out a square. If the square is very small you will need a lot of accuracy, but if the square is large the measurement can be done more crudely.

Let's take the case of a triangle on a plane. The sum of the angles is 180 degrees. Our friend on the sphere can find triangles that are very peculiar. He can, for example, find triangles which have *three right angles*. Yes indeed! One is shown in Figure 6-10. Suppose our bug starts at the north pole and makes a straight line all the way down to the equator. Then he makes a right angle and another perfect straight line the same length. Then he does it again. For the very special length he has chosen he gets right back to his starting point, and also meets the first line with a right angle. So there is no doubt that for him this triangle has three right angles, or 270 degrees in the sum. It turns out that for him the sum of the angles of the triangle is *always* greater than 180 degrees. In fact, the excess (for the special case shown, the extra 90 degrees) is proportional to how much area the triangle has. If a triangle on a sphere is very small, its angles add up to very nearly 180 degrees, only a little bit over. As the triangle gets bigger the discrepancy goes up. The bug on the hot plate would discover similar difficulties with his triangles.

Let's look next at what our other bugs find out about circles. They make circles and measure their circumferences. For example,

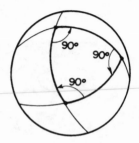

Figure 6-10. On a sphere a "triangle" can have three 90° angles.

Curved Space

the bug on the sphere might make a circle like the one shown in Figure 6-11. And he would discover that the circumference is *less* than 2π times the radius. (You can see that because from the wisdom of our three-dimensional view it is obvious that what he calls the "radius" is a curve which is *longer* than the true radius of the circle.) Suppose that the bug on the sphere had read Euclid, and decided to predict a radius by dividing the circumference C by 2π, taking

$$r_{\text{pred}} = \frac{C}{2\pi}.$$ (6.1)

Then he would find that the measured radius was larger than the predicted radius. Pursuing the subject, he might define the difference to be the "excess radius," and write

$$r_{\text{meas}} - r_{\text{pred}} = r_{\text{excess}},$$ (6.2)

and study how the excess radius effect depended on the size of the circle.

Our bug on the hot plate would discover a similar phenomenon. Suppose he was to draw a circle centered at the cold spot on the plate as in Figure 6-12. If we were to watch him as he makes the circle we would notice that his rulers are short near the center and get longer as they are moved outward—although the bug doesn't know it, of course. When he measures the circumference the ruler is

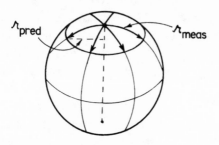

Figure 6-11. Making a circle on a sphere.

SIX NOT-SO-EASY PIECES

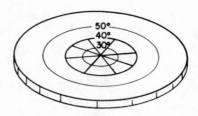

Figure 6-12. Making a circle on the hot plate.

long all the time, so he, too, finds out that the measured radius is longer than the predicted radius, $C/2\pi$. The hot-plate bug also finds an "excess radius effect." And again the size of the effect depends on the radius of the circle.

We will *define* a "curved space" as one in which these types of geometrical errors occur: The sum of the angles of a triangle is different from 180 degrees; the circumference of a circle divided by 2π is not equal to the radius; the rule for making a square doesn't give a closed figure. You can think of others.

We have given two different examples of curved space: the sphere and the hot plate. But it is interesting that if we choose the right temperature variation as a function of distance on the hot plate, the two *geometries* will be exactly the same. It is rather amusing. We can make the bug on the hot plate get exactly the same answers as the bug on the ball. For those who like geometry and geometrical problems we'll tell you how it can be done. If you assume that the length of the rulers (as determined by the temperature) goes in proportion to one plus some constant times the square of the distance away from the origin, then you will find that the geometry of that hot plate is exactly the same in all details* as the geometry of the sphere.

There are, of course, other kinds of geometry. We could ask about the geometry of a bug who lived on a pear, namely something which has a sharper curvature in one place and a weaker curvature

---

* except for the one point at infinity.

Curved Space

in the other place, so that the excess in angles in triangles is more severe when he makes little triangles in one part of his world than when he makes them in another part. In other words, the curvature of a space can vary from place to place. That's just a generalization of the idea. It can also be imitated by a suitable distribution of temperature on a hot plate.

We may also point out that the results could come out with the opposite kind of discrepancies. You could find out, for example, that all triangles when they are made too large have the sum of their angles *less* than 180 degrees. That may sound impossible, but it isn't at all. First of all, we could have a hot plate with the temperature decreasing with the distance from the center. Then all the effects would be reversed. But we can also do it purely geometrically by looking at the two-dimensional geometry of the surface of a saddle. Imagine a saddle-shaped surface like the one sketched in Figure 6-13. Now draw a "circle" on the surface, defined as the locus of all points the same distance from a center. This circle is a curve that oscillates up and down with a scallop effect. So its circumference is larger than you would expect from calculating $2\pi r$. So $C/2\pi$ is now less than r. The "excess radius" would be negative.

Spheres and pears and such are all surfaces of *positive* curvatures; and the others are called surfaces of *negative* curvature. In general, a two-dimensional world will have a curvature which varies from place to place and may be positive in some places and negative in

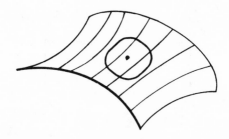

Figure 6-13. A "circle" on a saddle-shaped surface.

SIX NOT-SO-EASY PIECES

other places. In general, we mean by a curved space simply one in which the rules of Euclidean geometry break down with one sign of discrepancy or the other. The amount of curvature—defined, say, by the excess radius—may vary from place to place.

We might point out that, from our definition of curvature, a cylinder is, surprisingly enough, not curved. If a bug lived on a cylinder, as shown in Figure 6-14, he would find out that triangles, squares, and circles would all have the same behavior they have on a plane. This is easy to see, by just thinking about how all the figures will look if the cylinder is unrolled onto a plane. Then all the geometrical figures can be made to correspond exactly to the way they are in a plane. So there is no way for a bug living on a cylinder (assuming that he doesn't go all the way around, but just makes local measurements) to discover that his space is curved. In our technical sense, then, we consider that his space is *not* curved. What we want to talk about is more precisely called *intrinsic* curvature; that is, a curvature which can be found by measurements only in a local region. (A cylinder has no intrinsic curvature.) This was the

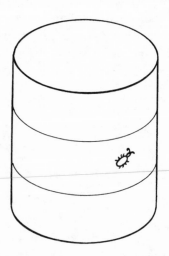

Figure 6-14. A two-dimensional space with zero intrinsic curvature.

Curved Space

sense intended by Einstein when he said that our space is curved. But we as yet only have defined a curved space in two dimensions; we must go onward to see what the idea might mean in three dimensions.

6-2 Curvature in three-dimensional space

We live in three-dimensional space and we are going to consider the idea that three-dimensional space is curved. You say, "But how can you imagine it being bent in any direction?" Well, we can't imagine space being bent in any direction because our imagination isn't good enough. (Perhaps it's just as well that we can't imagine too much, so that we don't get too free of the real world.) But we can still *define* a curvature without getting out of our three-dimensional world. All we have been talking about in two dimensions was simply an exercise to show how we could get a definition of curvature which didn't require that we be able to "look in" from the outside.

We can determine whether our world is curved or not in a way quite analogous to the one used by the gentlemen who live on the sphere and on the hot plate. We may not be able to distinguish between two such cases but we certainly can distinguish those cases from the flat space, the ordinary plane. How? Easy enough: We lay out a triangle and measure the angles. Or we make a great big circle and measure the circumference and the radius. Or we try to lay out some accurate squares, or try to make a cube. In each case we test whether the laws of geometry work. If they don't work, we say that our space is curved. If we lay out a big triangle and the sum of its angles exceeds 180 degrees, we can say our space is curved. Or if the measured radius of a circle is not equal to its circumference over 2π, we can say our space is curved.

You will notice that in three dimensions the situation can be much more complicated than in two. At any one place in two dimensions there is a certain amount of curvature. But in three dimensions there can be *several components* to the curvature. If we

SIX NOT-SO-EASY PIECES

lay out a triangle in some plane, we may get a different answer than if we orient the plane of the triangle in a different way. Or take the example of a circle. Suppose we draw a circle and measure the radius and it doesn't check with $C/2\pi$ so that there is some excess radius. Now we draw another circle at right angles—as in Figure 6-15. There's no need for the excess to be exactly the same for both circles. In fact, there might be a positive excess for a circle in one plane, and a defect (negative excess) for a circle in the other plane.

Perhaps you are thinking of a better idea: Can't we get around all of these components by using a *sphere* in three dimensions? We can specify a sphere by taking all the points that are the same distance from a given point in space. Then we can measure the surface area by laying out a fine scale rectangular grid on the surface of the sphere and adding up all the bits of area. According to Euclid the total area A is supposed to be 4π times the square of the radius; so we can define a "predicted radius" as $\sqrt{A/4\pi}$. But we can also measure the radius directly by digging a hole to the center and measuring the distance. Again, we can take the measured radius minus the predicted radius and call the difference the radius excess.

$$r_{\text{excess}} = r_{\text{meas}} - \left(\frac{\text{measured area}}{4\pi}\right)^{1/2},$$

which would be a perfectly satisfactory measure of the curvature.

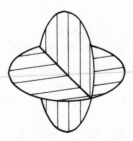

Figure 6-15. The excess radius may be different for circles with different orientations.

It has the great advantage that it doesn't depend upon how we orient a triangle or a circle.

But the excess radius of a sphere also has a disadvantage; it doesn't completely characterize the space. It gives what is called the *mean curvature* of the three-dimensional world, since there is an averaging effect over the various curvatures. Since it is an average, however, it does not solve completely the problem of defining the geometry. If you know only this number you can't predict all properties of the geometry of the space, because you can't tell what would happen with circles of different orientation. The complete definition requires the specification of six "curvature numbers" at each point. Of course the mathematicians know how to write all those numbers. You can read someday in a mathematics book how to write them all in a high-class and elegant form, but it is first a good idea to know in a rough way what it is that you are trying to write about. For most of our purposes the average curvature will be enough.*

6-3 Our space is curved

Now comes the main question. Is it true? That is, is the actual physical three-dimensional space we live in curved? Once we have enough imagination to realize the possibility that space might be curved, the human mind naturally gets curious about whether the real world is curved or not. People have made direct geometrical measurements to try to find out, and haven't found any deviations.

---

* We should mention one additional point for completeness. If you want to carry the hot-plate model of curved space over into three dimensions you must imagine that the length of the ruler depends not only on where you put it, but also on which orientation the ruler has when it is laid down. It is a generalization of the simple case in which the length of the ruler depends on where it is, but is the same if set north-south, or east-west, or up-down. This generalization is needed if you want to represent a three-dimensional space with any arbitrary geometry with such a model, although it happens not to be necessary for two dimensions.

On the other hand, by arguments about gravitation, Einstein discovered that space *is* curved, and we'd like to tell you what Einstein's law is for the amount of curvature, and also tell you a little bit about how he found out about it.

Einstein said that space is curved and that matter is the source of the curvature. (Matter is also the source of gravitation, so gravity is related to the curvature—but that will come later in the chapter.) Let us suppose, to make things a little easier, that the matter is distributed continuously with some density, which may vary, however, as much as you want from place to place.* The rule that Einstein gave for the curvature is the following: If there is a region of space with matter in it and we take a sphere small enough that the density ρ of matter inside it is effectively constant, then the *radius excess* for the sphere is proportional to the mass inside the sphere. Using the definition of excess radius, we have

$$\text{Radius excess} = \sqrt{\frac{A}{4\pi}} - r_{\text{meas}} = \frac{G}{3c^2} \cdot M. \qquad (6.3)$$

Here, G is the gravitational constant (of Newton's theory), c is the velocity of light, and $M = 4\pi\rho r^3/3$ is the mass of the matter inside the sphere. This is Einstein's law for the mean curvature of space.

Suppose we take the earth as an example and forget that the density varies from point to point—so we won't have to do any integrals. Suppose we were to measure the surface of the earth very carefully, and then dig a hole to the center and measure the radius. From the surface area we could calculate the predicted radius we would get from setting the area equal to $4\pi r^2$. When we compared the predicted radius with the actual radius, we would find that the actual radius exceeded the predicted radius by the amount given in Eq. (6.3). The constant $G/3c^2$ is about 2.5×10^{-29} cm per gram, so for each gram of material the measured radius is off by 2.5×10^{-29} cm. Putting in the mass of the earth, which is about 6×10^{27} grams, it

* Nobody—not even Einstein—knows how to do it if mass comes concentrated at points.

Curved Space

turns out that the earth has 1.5 millimeters more radius than it should have for its surface area.* Doing the same calculation for the sun, you find that the sun's radius is one-half a kilometer too long.

You should note that the law says that the *average* curvature *above* the surface area of the earth is zero. But that does *not* mean that all the components of the curvature are zero. There may still be—and, in fact, there is—some curvature above the earth. For a circle in a plane there will be an excess radius of one sign for some orientations and of the opposite sign for other orientations. It just turns out that the average over a sphere is zero when there is no mass *inside* it. Incidentally, it turns out that there is a relation between the various components of the curvature and the *variation* of the average curvature from place to place. So if you know the average curvature everywhere, you can figure out the details of the curvature at each place. The average curvature above the earth varies with altitude, so the space there is curved. And it is that curvature that we see as a gravitational force.

Suppose we have a bug on a plane, and suppose that the "plane" has little pimples in the surface. Wherever there is a pimple the bug would conclude that his space has little local regions of curvature. We have the same thing in three dimensions. Wherever there is a lump of matter, our three-dimensional space has a local curvature— a kind of three-dimensional pimple.

If we make a lot of bumps on a plane there might be an overall curvature besides all the pimples—the surface might become like a ball. It would be interesting to know whether our space has a net average curvature as well as the local pimples due to the lumps of matter like the earth and the sun. The astrophysicists have been trying to answer that question by making measurements of galaxies at very large distances. For example, if the number of galaxies we see in a spherical shell at a large distance is different from what we

---

* Approximately, because the density is not independent of radius as we are assuming.

would expect from our knowledge of the radius of the shell, we would have a measure of the excess radius of a tremendously large sphere. From such measurements it is hoped to find out whether our whole universe is flat on the average, or round—whether it is "closed," like a sphere, or "open" like a plane. You may have heard about the debates that are going on about this subject. There are debates because the astronomical measurements are still completely inconclusive; the experimental data are not precise enough to give a definite answer. Unfortunately, we don't have the slightest idea about the overall curvature of our universe on a large scale.

6-4 Geometry in space-time

Now we have to talk about time. As you know from the special theory of relativity, measurements of space and measurements of time are interrelated. And it would be kind of crazy to have something happening to the space, without the time being involved in the same thing. You will remember that the measurement of time depends on the speed at which you move. For instance, if we watch a guy going by in a space ship we see that things happen more slowly for him than for us. Let's say he takes off on a trip and returns in 100 seconds flat *by our watches*; his watch might say that he had been gone for only 95 seconds. In comparison with ours, his watch—and all other processes, like his heart beat—have been running slow.

Now let's consider an interesting problem. Suppose you are the one in the space ship. We ask you to start off at a given signal and return to your starting place just in time to catch a later signal—at, say, exactly 100 seconds later according to *our* clock. And you are also asked to make the trip in such a way that *your* watch will show the *longest possible* elapsed time. How should you move? You should stand still. If you move at all your watch will read less than 100 seconds when you get back.

Curved Space

Suppose, however, we change the problem a little. Suppose we ask you to start at Point A on a given signal and go to point B (both fixed relative to us), and to do it in such a way that you arrive back just at the time of a second signal (say 100 seconds later according to our fixed clock). Again you are asked to make the trip in the way that lets you arrive with the latest possible reading on your watch. How would you do it? For which path and schedule will *your* watch show the greatest elapsed time when you arrive? The answer is that you will spend the longest time from *your* point of view if you make the trip by going at a uniform speed along a straight line. Reason: Any extra motions and any extra-high speeds will make your clock go slower. (Since the time deviations depend on the *square* of the velocity, what you lose by going extra fast at one place you can never make up by going extra slowly in another place.)

The point of all this is that we can use the idea to define "a straight line" in space-time. The analog of a straight line in space is for space-time a *motion* at uniform velocity in a constant direction.

The curve of shortest distance in space corresponds in space-time not to the path of shortest time, but to the one of *longest* time, because of the funny things that happen to signs of the *t*-terms in relativity. "Straight-line" motion—the analog of "uniform velocity along a straight line"—is then that motion which takes a watch from one place at one time to another place at another time in the way that gives the longest time reading for the watch. This will be our definition for the analog of a straight line in space-time.

6-5 Gravity and the principle of equivalence

Now we are ready to discuss the laws of gravitation. Einstein was trying to generate a theory of gravitation that would fit with the relativity theory that he had developed earlier. He was struggling along until he latched onto one important principle which guided him into getting the correct laws. That principle is based on the idea that when a thing is falling freely everything inside it seems weightless. For example, a satellite in orbit is falling freely in the earth's gravity, and an astronaut in it feels weightless. This idea, when stated with greater precision, is called *Einstein's principle of equivalence*. It depends on the fact that all objects fall with exactly the same acceleration no matter what their mass, or what they are made of. If we have a spaceship that is "coasting"—so it's in a free fall—and there is a man inside, then the laws governing the fall of the man and the ship are the same. So if he puts himself in the middle of the ship he will stay there. He doesn't fall *with respect to the ship*. That's what we mean when we say he is "weightless."

Now suppose you are in a rocket ship which is accelerating. Accelerating with respect to what? Let's just say that its engines are on and generating a thrust so that it is not coasting in a free fall. Also imagine that you are way out in empty space so that there are practically no gravitational forces on the ship. If the ship is accelerating with "1g" you will be able to stand on the "floor" and will feel your normal weight. Also if you let go of a ball, it will "fall" toward the floor. Why? Because the ship is accelerating "upward," but the ball has no forces on it, so it will not accelerate; it will get left behind. Inside the ship the ball will appear to have a downward acceleration of "1g."

Now let's compare that with the situation in a spaceship sitting at rest on the surface of the earth. *Everything is the same!* You would be pressed toward the floor, a ball would fall with an acceleration of 1g, and so on. In fact, how could you tell inside a space ship whether you are sitting on the earth or are accelerating in free space? According to Einstein's equivalence principle there is no way to tell if you only make measurements of what happens to things inside!

To be strictly correct, that is true only for one point inside the ship. The gravitational field of the earth is not precisely uniform, so a freely falling ball has a slightly different acceleration at different places—the direction changes and the magnitude changes. But if we imagine a strictly uniform gravitational field, it is completely imitated in every respect by a system with a constant acceleration. That is the basis of the principle of equivalence.

Curved Space

6-6 The speed of clocks in a gravitational field

Now we want to use the principle of equivalence for figuring out a strange thing that happens in a gravitational field. We'll show you something that happens in a rocket ship which you probably wouldn't have expected to happen in a gravitational field. Suppose we put a clock at the "head" of the rocket ship—that is, at the "front" end—and we put another identical clock at the "tail," as in Figure 6-16. Let's call the two clocks A and B. If we compare these two clocks when the ship is accelerating, the clock at the head seems

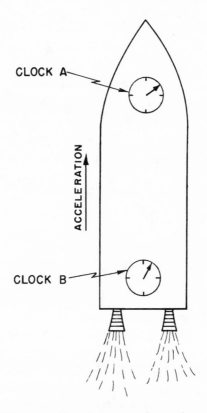

Figure 6-16. An accelerating rocket ship with two clocks.

to run fast relative to the one at the tail. To see that, imagine that the front clock emits a flash of light each second, and that you are sitting at the tail comparing the arrival of the light flashes with the ticks of clock B. Let's say that the rocket is in the position *a* of Figure 6-17 when clock A emits a flash, and at the position *b* when the flash arrives at clock B. Later on the ship will be at position *c* when the clock A emits its next flash, and at position *d* when you see it arrive at clock B.

The first flash travels the distance L_1 and the second flash travels the shorter distance L_2. It is a shorter distance because the ship is accelerating and has a higher speed at the time of the second flash. You can see, then, that if the two flashes were emitted from clock A one second apart, they would arrive at clock B with a separation somewhat less than one second, since the second flash doesn't spend as much time on the way. The same thing will also happen for all the later flashes. So if you were sitting in the tail you would conclude that clock A was running faster than clock B. If you were to do the same thing in reverse—letting clock B emit light and observing it at clock A—you would conclude that B was running

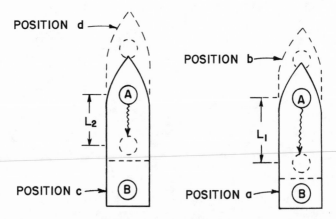

Figure 6-17. A clock at the head of an accelerating rocket ship appears to run faster than a clock at the tail.

Curved Space

slower than *A*. Everything fits together and there is nothing myste-rious about it all.

But now let's think of the rocket ship at rest in the earth's gravity. *The same thing happens.* If you sit on the floor with one clock and watch another one which is sitting on a high shelf, it will appear to run faster than the one on the floor! You say, "But that is wrong. The times should be the same. With no acceleration there's no reason for the clocks to appear to be out of step." But they must if the principle of equivalence is right. And Einstein insisted that the principle *was* right, and went courageously and correctly ahead. He proposed that clocks at different places in a gravita-tional field must appear to run at different speeds. But if one always *appears* to be running at a different speed with respect to the other, then so far as the first is concerned the other *is* running at a different rate.

But now you see we have the analog for clocks of the hot ruler we were talking about earlier, when we had the bug on a hot plate. We imagined that rulers and bugs and everything changed lengths in the same way at various temperatures so they could never tell that their measuring sticks were changing as they moved around on the hot plate. It's the same with clocks in a gravitational field. Every clock we put at a higher level is seen to go faster. Heartbeats go faster, all processes run faster.

If they didn't you would be able to tell the difference between a gravitational field and an accelerating reference system. The idea that time can vary from place to place is a difficult one, but it is the idea Einstein used, and it is correct—believe it or not.

Using the principle of equivalence we can figure out how much the speed of a clock changes with height in a gravitational field. We just work out the apparent discrepancy between the two clocks in the accelerating rocket ship. The easiest way to do this is to use the result we found in Chapter 34 of Vol. I* for the Doppler effect. There, we found—see Eq. (34.14)*—that if v is the *relative* velocity

* of the original *Lectures on Physics.*

of a source and a receiver, the *received* frequency ω is related to the *emitted* frequency ω_0 by

$$\omega = \omega_0 \, \frac{1 + v/c}{\sqrt{1 - v^2/c^2}} \, . \qquad (6.4)$$

Now if we think of the accelerating rocket ship in Figure 6-17 the emitter and receiver are moving with equal velocities at any one instant. But in the time that it takes the light signals to go from clock A to clock B the ship has accelerated. It has, in fact, picked up the additional velocity gt, where g is the acceleration and t is time it takes light to travel the distance H from A to B. This time is very nearly H/c. So when the signals arrive at B, the ship has increased its velocity by gH/c. The receiver always has this velocity *with respect to the emitter* at the instant the signal left it. So this is the velocity we should use in the Doppler shift formula, Eq. (6.4). Assuming that the acceleration and the length of the ship are small enough that this velocity is much smaller than c, we can neglect the term in v^2/c^2. We have that

$$\omega = \omega_0 \left(1 + \frac{gH}{c^2} \right). \qquad (6.5)$$

So for the two clocks in the space ship we have the relation

$$\text{(Rate at the receiver)} = \text{(Rate of emission)} \left(1 + \frac{gH}{c^2} \right), \qquad (6.6)$$

where H is the height of the emitter *above* the receiver.

From the equivalence principle the same result must hold for two clocks separated by the height H in a gravitational field with the free fall acceleration g.

This is such an important idea we would like to demonstrate that it also follows from another law of physics—from the conservation of energy. We know that the gravitational force on an object is proportional to its mass M, which is related to its total internal energy E by $M = E/c^2$. For instance, the masses of nuclei determined from the *energies* of nuclear reactions which transmute one nucleus into another agree with the masses obtained from atomic *weights*.

Curved Space

Now think of an atom which has a lowest energy state of total energy E_0 and a higher energy state E_1, and which can go from the state E_1 to the state E_0 by emitting light. The frequency ω of the light will be given by

$$\hbar\omega = E_1 - E_0. \tag{6.7}$$

Now suppose we have such an atom in the state E_1 sitting on the floor, and we carry it from the floor to the height H. To do that we must do some work in carrying the mass $m_1 = E_1/c^2$ up against the gravitational force. The amount of work done is

$$\frac{E_1}{c^2} gH. \tag{6.8}$$

Then we let the atom emit a photon and go into the lower energy state E_0. Afterward we carry the atom back to the floor. On the return trip the mass is E_0/c^2; we get back the energy

$$\frac{E_0}{c^2} gH, \tag{6.9}$$

so we have done a net amount of work equal to

$$\Delta U = \frac{E_1 - E_0}{c^2} gH. \tag{6.10}$$

When the atom emitted the photon it gave up the energy $E_1 - E_0$. Now suppose that the photon happened to go down to the floor and be absorbed. How much energy would it deliver there? You might at first think that it would deliver just the energy $E_1 - E_0$. But that can't be right if energy is conserved, as you can see from the following argument. We started with the energy E_1 at the floor. When we finish, the energy at the floor level is the energy E_0 of the atom in its lower state plus the energy E_{ph} received from the photon. In the meantime we have had to supply the additional energy ΔU of Eq. (6.10). If energy is conserved, the energy we end up with at the floor must be greater than we started with by just the work we have done. Namely, we must have that

SIX NOT-SO-EASY PIECES

$$E_{ph} + E_0 = E_1 + \Delta U,$$

or   (6.11)

$$E_{ph} = (E_1 - E_0) + \Delta U.$$

It must be that the photon does *not* arrive at the floor with just the energy $E_1 - E_0$ it started with, but with a *little more energy*. Otherwise some energy would have been lost. If we substitute in Eq. (6.11) the ΔU we got in Eq. (6.10), we get that the photon arrives at the floor with the energy

$$E_{ph} = (E_1 - E_0)\left(1 + \frac{gH}{c^2}\right).$$ (6.12)

But a photon of energy E_{ph} has the frequency $\omega = E_{ph}/\hbar$. Calling the frequency of the *emitted* photon ω_0—which is by Eq. (6.7) equal to $(E_1 - E_0)/\hbar$—our result in Eq. (6.12) gives again the relation of (6.5) between the frequency of the photon when it is absorbed on the floor and the frequency with which it was emitted.

The same result can be obtained in still another way. A photon of frequency ω_0 has the energy $E_0 = \hbar\omega_0$. Since the energy E_0 has the gravitational mass E_0/c^2 the photon has a mass (*not* rest mass) $\hbar\omega_0/c^2$, and is "attracted" by the earth. In falling the distance H it will gain an additional energy $(\hbar\omega_0/c^2)gH$, so it arrives with the energy

$$E = \hbar w_0\left(1 + \frac{gH}{c^2}\right).$$

But its frequency after the fall is $E/\hbar$, giving again the result in Eq. (6.5). Our ideas about relativity, quantum physics, and energy conservation all fit together only if Einstein's predictions about clocks in a gravitational field are right. The frequency changes we are talking about are normally very small. For instance, for an altitude difference of 20 meters at the earth's surface the frequency difference is only about two parts in 10^{15}. However, just such a change has recently been found experimentally using the Mössbauer effect.* Einstein was perfectly correct.

* R. V. Pound and G. A. Rebka, Jr., *Physical Review Letters* Vol. 4, p. 337 (1960).

Curved Space

6-7 *The curvature of space-time*

Now we want to relate what we have just been talking about to the idea of curved space-time. We have already pointed out that if the time goes at different rates in different places, it is analogous to the curved space of the hot plate. But it is more than an analogy; it means that space-time *is* curved. Let's try to do some geometry in space-time. That may at first sound peculiar, but we have often made diagrams of space-time with distance plotted along one axis and time along the other. Suppose we try to make a rectangle in space-time. We begin by plotting a graph of height H versus t as in Figure 6-18(a). To make the base of our rectangle we take an object which is *at rest* at the height H_1 and follow its world line for 100 seconds. We get the line BD in part (b) of the figure which is parallel to the t-axis. Now let's take another object which is 100 feet above the first one at $t = 0$. It starts at the point A in Figure 6-18(c). Now we follow its world line for 100 seconds as measured by a clock at A. The object goes from A to C, as shown in part (d) of the figure. But notice that since time goes at a different rate at the two heights—we are assuming that there is a gravitational field—the two points C and D are not simultaneous. If we try to complete the square by drawing a line to the point C' which is 100 feet above D at the same time, as in Figure 6-18(e), the pieces don't fit. And that's what we mean when we say that space-time is curved.

6-8 *Motion in curved space-time*

Let's consider an interesting little puzzle. We have two identical clocks, A and B, sitting together on the surface of the earth as in Figure 6-19. Now we lift clock A to some height H, hold it there awhile, and return it to the ground so that it arrives at just the instant when clock B has advanced by 100 seconds. Then clock A will read something like 107 seconds, because it was running faster when it was up in the air. Now here is the puzzle. How should we

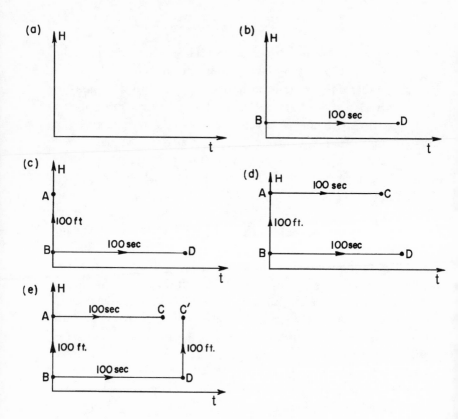

Figure 6-18. Trying to make a rectangle in space-time.

move clock *A* so that it reads the latest possible time—always assuming that it returns when *B* reads 100 seconds? You say, "That's easy. Just take *A* as high as you can. Then it will run as fast as possible, and be the latest when you return." Wrong. You forgot something—we've only got 100 seconds to go up and back. If we go very high, we have to go very fast to get there and back in 100 seconds. And you mustn't forget the effect of special relativity which causes moving clocks to *slow down* by the factor $\sqrt{1 - v^2/c^2}$. This relativity effect works in the direction of making clock *A* read

Curved Space

Combining the two terms in (6.13) and (6.14) we have that

$$\Delta\omega = \frac{\omega_0}{c^2}\left(gH - \frac{v^2}{2}\right).$$ (6.15)

Such a frequency shift of our moving clock means that if we measure a time dt on a fixed clock, the moving clock will register the time

$$dt\left[1 + \left(\frac{gH}{c^2} - \frac{v^2}{2c^2}\right)\right].$$ (6.16)

The total time excess over the trajectory is the integral of the extra term with respect to time, namely

$$\frac{1}{c^2}\int\left(gH - \frac{v^2}{2}\right)\,dt,$$ (6.17)

which is supposed to be a maximum.

The term gH is just the gravitational potential ϕ. Suppose we multiply the whole thing by a constant factor $-mc^2$, where m is the mass of the object. The constant won't change the condition for the maximum, but the minus sign will just change the maximum to a minimum. Equation (6.16) then says that the object will move so that

$$\int\left(\frac{mv^2}{2} - m\phi\right)dt = \text{a minimum}.$$ (6.18)

But now the integrand is just the difference of the kinetic and potential energies. And if you look in Chapter 19 of Volume II* you will see that when we discussed the principle of least action we showed that Newton's laws for an object in any potential could be written exactly in the form of Eq. (6.18).

6-9 Einstein's theory of gravitation

Einstein's form of the equations of motion—that the proper time should be a maximum in curved space-time—gives the same results

* of the original *Lectures on Physics*.

as Newton's laws for low velocities. As he was circling around the earth, Gordon Cooper's watch was reading later than it would have in any other path you could have imagined for his satellite.*

So the law of gravitation can be stated in terms of the ideas of the geometry of space-time in this remarkable way. The particles always take the longest proper time—in space-time a quantity analogous to the "shortest distance." That's the law of motion in a gravitational field. The great advantage of putting it this way is that the law doesn't depend on any coordinates, or any other way of defining the situation.

Now let's summarize what we have done. We have given you two laws for gravity:

(1) How the geometry of space-time changes when matter is present—namely, that the curvature expressed in terms of the excess radius is proportional to the mass inside a sphere, Eq. (6.3).

(2) How objects move if there are only gravitational forces—namely, that objects move so that their proper time between two end conditions is a maximum.

Those two laws correspond to similar pairs of laws we have seen earlier. We originally described motion in a gravitational field in terms of Newton's inverse square law of gravitation and his laws of motion. Now laws (1) and (2) take their places. Our new pair of laws also correspond to what we have seen in electrodynamics. There we had our law—the set of Maxwell's equations—which determines the fields produced by charges. It tells how the character of "space" is changed by the presence of charged matter, which is what law (1) does for gravity. In addition, we had a law about how

---

* Strictly speaking it is only a *local* maximum. We should have said that the proper time is larger than for any *nearby* path. For example, the proper time on an elliptical orbit around the earth need not be longer than on a ballistic path of an object which is shot to a great height and falls back down.

Curved Space

particles move in the given fields—$d(m\mathbf{v})/dt = q(\mathbf{E} + \mathbf{v} \times \mathbf{B})$. This, for gravity, is done by law (2).

In the laws (1) and (2) you have a precise statement of Einstein's theory of gravitation—although you will usually find it stated in a more complicated mathematical form. We should, however, make one further addition. Just as time scales change from place to place in a gravitational field, so do the length scales. Rulers change lengths as you move around. It is impossible with space and time so intimately mixed to have something happen with time that isn't in some way reflected in space. Take even the simplest example: You are riding past the earth. What is "*time*" from *your* point of view is partly space from *our* point of view. So there must also be changes in space. It is the entire *space-time* which is distorted by the presence of matter, and this is more complicated than a change only in time scale. However, the rule that we gave in Eq. (6-3) is enough to determine completely all the laws of gravitation, provided that it is understood that this rule about the curvature of space applies not only from one man's point of view but is true for everybody. Somebody riding by a mass of material sees a different mass content because of the kinetic energy he calculates for its motion past him, and he must include the mass corresponding to that energy. The theory must be arranged so that everybody—no matter how he moves—will, when he draws a sphere, find that the excess radius is $G/3c^2$ times the total mass (or, better, $G/3c^4$ times the total energy content) inside the sphere. That this law—law (1)—should be true in any moving system is one of the great laws of gravitation, called *Einstein's field equation*. The other great law is (2)—that things must move so that the proper time is a maximum—and is called *Einstein's equation of motion*.

To write these laws in a complete algebraic form, to compare them with Newton's laws, or to relate them to electrodynamics is difficult mathematically. But it is the way our most complete laws of the physics of gravity look today.

Although they gave a result in agreement with Newton's mechanics for the simple example we considered, they do not always do so. The three discrepancies first derived by Einstein have been experimentally confirmed: The orbit of Mercury is not a fixed ellipse; starlight passing near the sun is deflected twice as much as you would think; and the rates of clocks depend on their location in a gravitational field. Whenever the predictions of Einstein have been found to differ from the ideas of Newtonian mechanics, Nature has chosen Einstein's.

Let's summarize everything that we have said in the following way. First, time and distance rates depend on the place in space you measure them and on the time. This is equivalent to the statement that space-time is curved. From the measured area of a sphere we can define a predicted radius, $\sqrt{A/4\pi}$, but the actual measured radius will have an excess over this which is proportional (the constant is G/c^2) to the total mass contained inside the sphere. This fixes the exact degree of the curvature of space-time. And the curvature must be the same no matter who is looking at the matter or how it is moving. Second, particles move on "straight lines" (trajectories of maximum proper time) in this curved space-time. This is the content of Einstein's formulation of the laws of gravitation.

Index

Acceleration vector, 15–18
Affective future, 100, 101
Affective past, 100
Angular momentum, conservation of, 29
Angular orientation, and physical laws, 6–8
Antiatoms, 44
Antimatter, 43–46
Antineutrons, 44
Antiparticles, 44
Antiprotons ($\bar{p}$), 44
Atomic bomb, energy of, 89–90
Atoms
    anti-atoms of, 44
    conservation of energy and, 134–136
    mass at rest, 89–90
    reversibility in time and, 29
    scale of, 27
Axial vectors, 35–38

Beta decay, 40–43, 45
Billiard bills, collision of, 51–52

Centrifugal force, 76
Challenger (space shuttle), *xix, xvi*
Clocks
    slowing in moving system, 59–62
    speed in gravitational field, 131–136, 140–141
Cobalt, 41–42, 45
Collision, analysis of, 83–88, 105
    inelastic, 87–88, 89

Conservation laws
    energy, conservation of. *See* Energy, conservation of
    four-vectors, 106–107
    mass, conservation of, 88, 89
    momentum, conservation of. *See* Momentum, conservation of
    parity, conservation of, 40–43
    symmetry and, 29–30
Conspiracy as law of nature, complete, 58
Current time (space-time region), 99–101
Curved space (curvature), 111–144
    defined, 117, 120, 122, 123, 125
    Einstein's law for, 126–128
    geometry of, 117–125, 128–129
    intrinsic, 122–123
    mean, 125, 126–127
    negative, 121–122, 124
    numbers, 125
    positive, 121–122, 124
    space-time geometry and, 128–129
    three-dimensional, 123–125
    two-dimensional, 111–123
Cylinders, curvature of, 122–123

Directed quantities, 10
Direction
    effect of, 5. *See also* Rotation
    of vectors, 20–21
Displacement, 10
Doppler effect, 109, 133–134
Dot products (vectors), 19–21, 107–108
Dyson, Freeman, *x*

Earth, velocity of, 53, 54–59
Einstein, Albert
  curved space-time laws, *xiv, xv,*
    111, 123, 126–128
  equivalence principle, 129–136
  field equation, *xvi,* 143–144
  gravitation theory, 129, 141–144
  Lorentz transformation and, 54
  motion, equation of, 143–144
  relativity theories, *ix, xii, xiii, xv–*
    xvi, 50, 55, 66, 68–69, 73, 79
Electrical charge, conservation of,
  30
Electricity
  laws of, 40
  theories of relativity and, 52
Electrodynamics, Maxwell's laws of,
  xii–xiii, 52–54, 75, 142
Electron volts, energy measured in,
  103
Electrons, 43–44, 71
Energy
  conservation of, 29, 89, 107, 134–
    136
  equivalence of mass and, *xvi,* 68–
    71, 102–103
  of photon, 108–109
  relativistic, *xvi,* 88–91
  as time component of vectors, 102–
    105
  total energy of particle, 91
  vector analysis, 20
  velocity, momentum and total, 91
Equivalence, Einstein's principle of,
  129–136
Euclid, 124
Event, in space-time, 95

Feynman, Richard, 153–154
  as scientist, *ix–xvi*
  as teacher, *ix, xix–xxiii*
Feynman Lectures on Physics, The, xi
  Feynman's Preface, *xxv–xxix*
  Special Preface, *xix–xxiii*

Force
  Newton's law on momentum and,
    66
  rotation and, 7–8
  vector analysis, 10, 11–12, 16
Foucault pendulum, 28, 76
Four-vector momentum, 105
Four-vectors, *xiv–xv,* 65–66, 102–109
Future (time-like interval), 99–100,
  101

Galilean transformation, 52, 53, 54
Galileo, *xii–xiii,* 27–28
Gamma(τ)-meson, 40
General Theory of Relativity, *xv,* 50
Geographical differences, effect on
    physical laws of, 25
Geometry
  of curved space, 117–125, 128–
    129
  of space-time. *See* Space-time
Gravitation, laws of
  change of scale, effect of, 27
  Einstein's theory of, 141–144
  equivalence principle and, 129–136
  Newton's laws of, 111, 142
  reflection symmetry in, 40
Gravitational field
  motion law in, 140, 141–144
  speed of clocks in, 131–136, 140–
    141

Huygens, Christian, 51

Inelastic collision, 87–88, 89
Inertia, 67, 88
Intervals, space-time, 97–99
Invariance. *See* Symmetry
Invariant. *See* Scalar

Kinetic energy, 20, 68–69, 88–91
K-mesons, 91

Least action, principle of, 141

Left-handedness, 31–39

Leighton, Robert B., *xi*

Light

distance measured by speed of, 98

photon of, 108–109

representation in space-time, 96, 98

velocity of, *xiii,* 52–53, 56–57, 75, 98, 100

wavelength of, variations in, 27

Lorentz, Hendrik Antoon, *xii, xiii, xv,* 58

Lorentz transformation, 26, 53–54

consequences of, 79–83

contraction, 63

laws of mechanics under, 66

as rotation in time and space, 65–66, 102, 105

simultaneity, 63–64

space-time relationships implied by, 93–95, 98

twin paradox, 77–79

Magnetic fields

cobalt experiment, 41–42, 45

electromagnetic field, Maxwell's equations of. *See* Electrodynamics, Maxwell's laws of

geographical differences, effect of, 25

poles and reflection symmetry, 36–38, 41–42, 45

Magnetism

laws of, 40

principle of relativity and, 52

Mass

of colliding objects, 83–88

conservation of, 88, 89

equivalence of energy and, *xvi,* 68–71, 102–103

in motion, 91

relativity of, *xvi,* 49–50, 66–68, 83–88

rest, 49, 89–90, 91

rest mass equaling zero, 108–109

temperature increase and, 68

variation with speed, 69

Matter, and antimatter, 43–46

Maxwell, James Clerk. *See* Electrodynamics, Maxwell's laws of

Mechanics, laws of

equations, 2–3

Lorentz transformation, effect of, 66

Michelson-Morley experiment, 54–59, 60, 63

Minkowski, Hermann, *xiv–xv*

Mirror reflections. *See* Reflection symmetry

Momentum, 10, 20

conservation of, 29, 52, 66, 85–88, 106–107

four-vector, 105

as function of velocity, 84

Newton's law on force and, 66

of photon, 108–109

speed, variation with, 66–67

three-vector, 105, 106

transverse, 82–83, 85–88, 105

vector components, 102–105

velocity, total energy and, 91

Mössbauer effect, 136

Motion. *See also* Velocity

absolute, 75–76

in curved space-time, 137–141

Einstein's equation of, 141, 143–144

of translation, uniform, 73

vertical component of, 82–83, 105

Motion, laws of, 50–51

gravitational field, law of motion in, 140–144

Newton's laws, 111

Mu-mesons (muons), 62, 78–79

Neutrons, 44

Newton's laws, 3, 4, 5, 8–9

Newton's laws (*continued*)
  force as rate of change of
    momentum, 66
  invariance of, 9, 16
  least action, principle of, 141
  of motion, 111
  principle of relativity and, *xii–xiii,*
    50–51, 54, 66–68, 75, 77, 79, 89
  Second Law, 49
  of universal gravitation, 111
  in vector notation, 15–18
Nuclear forces, 40, 46

Ordinary quantity, 10

Parity, conservation of, 40–43
  failure of, 42–43
Past (time-like interval), 99–100
Philosophers, and relativity, 73–77
Photons, 108–109, 135–136
Physics, laws of. *See also* specific laws,
    e.g. Conservation laws
  change of scale, effect of, 26–28, 39
  invariance of, 9, 16
  reversibility in time of, 28–29
  symmetry of, 2–3, 77
  vector notation, 16
Planck's constant, 108
Poincaré, Henri, *xii, xiii, xv,* 54, 58,
    73
Polar vectors, 35–38
Positrons, 43–44, 71
Protons, 44, 46

Quantum mechanics, *ix, xv,* 29, 30

Reflection symmetry, *xiv,* 30–35
  antimatter and, 45
  identification of right and left, 38–
    39
  polar and axial vectors, 35–38
Relativity, *vii, xii. See also* Lorentz
    transformation
  addition of velocities in, 80–82

energy, relativistic, *xvi,* 88–91
  General Theory of, *xv,* 50
  of mass, *xvi,* 49–50, 66–68, 83–88
  Michelson and Morley experiment,
    54–59, 60, 63
  momentum, law of conservation of,
    107
  philosophers and, 73–77
  principle of, 49–53, 73, 75
  relativistic dynamics, 66–68
  simultaneity, 63–64
  Special Theory of, *ix, xv,* 49–71,
    128
  time, transformation of, 59–62
Right-handedness, 31–39
Rotation
  Lorentz transformation, 65
  space-time measurements, 93–94,
    102
  symmetry and, *xiv,* 5–9, 16, 25, 28,
    29
  uniform, 76

Sands, Matthew, *xi, xxviii*
Scalar, 10, 19
Scalar function, 19
Scalar product, 18–21, 108
Scale, changes in, 26–28, 39
Schrödinger equation, 33–34
Schwinger, Julian, *xxii*
Seconds, measure of distance by, 98
Simultaneity, 63–64, 101
  failure at a distance, 64
Sound waves, speed of, 52
Space
  curved. *See* Curved space
  reflection in, 30–35
  rotation in, 25, 28, 29
  space-time compared, 99
  symmetry in, 24–29
Space-like interval, 99
Space-time, *xv,* 93–109
  curvature of, *xiv–xv,* 111, 137–141
  events in, 95

geometry of, *xiv–xv,* 93–97, 128–
129, 137, 142
intervals, 97–99
ordinary space compared, 99
past, present, and future in, 99–101
"straight line" in, 129
Special Theory of Relativity, *ix, xv,*
49–71, 128
Speed. *See* Velocity
Symmetry, *xiii–xiv,* 1–2, 23–47. *See
also* Vectors
antimatter, 43–46
broken, 46–47
conservation laws and, 29–30
conservation of parity and, 40–43
defined, 1, 24
operations, 23–24
of physical laws, 77
polar and axial vectors, 35–38
reflection. *See* Reflection symmetry
rotation of axes, *xiv,* 5–9, 16, 25,
28, 29
scale, effect of changes in, 26–28,
39
in space and time, 24–29
translational displacements, 2–5
under uniform velocity in straight
line, 25–26

Temperature
mass and, 68
as scalar, 10
Theta(Θ)-meson, 40
Three-dimensional curved space, 123–
125
Three-vectors, 102, 105, 106
Time. *See also* Space-time
curved, 111. *See also* Curved space
displacement in, 25
reversibility in, 28–29
symmetry in, 24–29
transformation of, 59–62
vector component, 65–66, 102–105
Time-like interval, 99–100, 101

Tomanaga, Sin-Itero, *xxii*
Transformations. *See also* Galilean
transformation; Lorentz
transformation
four-vector momentum, 105
Translation, uniform motion of, 73
Translational displacements, 2–5
Twin paradox, 77–79
Two-dimensional curved space, 111–
123

Undirected quantity, 10
Unit vectors, 21
Upsilon(Y)-rays, 44
Uranium atom, 89–90

Vector algebra, 12–15
addition, 12–13, 15
four-vector, 106–109
multiplication, 13–14, 15
subtraction, 14, 15, 16–17
Vectors, *xiv,* 9–12
acceleration, 15–18
axial, 35–38
components of, 10–11
defined, 10
direction of, 20–21
four-vectors, *xiv–xv,* 65–66, 102–
109
Newton's laws in vector notation,
15–18
polar, 35–38
properties of, 11–21
scalar product of, 18–21
symbol to mark, 10–11
three-vectors, 102, 105, 106
time components, 65–66, 102–105
unit, 21
velocity, 14–15, 16–18, 102, 103
Velocity
absolute, 58, 74–76
addition of velocities in relativity,
80–82
direction as property of, 10

Velocity (*continued*)
  of earth, 53, 54–59
  of light, *xiii,* 52–53, 56–57, 75, 98,
    100
  mass and, 49–53, 69
  momentum and, 66–67, 84, 91
  of sound waves, 52
  total energy and, 91
  transformation of, 79–83
  uniform, 76

vectors, 14–15, 16–18, 102, 103
vertical component of, 82–83, 85–
    88, 105

Waves
  light, variations in length of, 27
  sound, speed of, 52
Weak decay, 40–43, 45
Weyl, Hermann, 1, 24
Work, 20, 88

About Richard Feynman

Born in 1918 in Brooklyn, Richard P. Feynman received his Ph.D. from Princeton in 1942. Despite his youth, he played an important part in the Manhattan Project at Los Alamos during World War II. Subsequently, he taught at Cornell and at the California Institute of Technology. In 1965 he received the Nobel Prize in Physics, along with Sin-Itero Tomanaga and Julian Schwinger, for his work in quantum electrodynamics.

Dr. Feynman won his Nobel Prize for successfully resolving problems with the theory of quantum electrodynamics. He also created a mathematical theory that accounts for the phenomenon of superfluidity in liquid helium. Thereafter, with Murray Gell-Mann, he did fundamental work in the area of weak interactions such as beta decay. In later years Feynman played a key role in the development of quark theory by putting forward his parton model of high energy proton collision processes.

Beyond these achievements, Dr. Feynman introduced basic new computational techniques and notations into physics, above all, the ubiquitous Feynman diagrams that, perhaps more than any other formalism in recent scientific history, have changed the way in which basic physical processes are conceptualized and calculated.

Feynman was a remarkably effective educator. Of all his numerous awards, he was especially proud of the Oersted Medal for Teaching which he won in 1972. *The Feynman Lectures on Physics*, originally published in 1963, were described by a reviewer in *Scientific American* as "tough, but nourishing and full of flavor. After 25 years it is *the* guide for teachers and for the best of beginning students." In order to increase the understanding of physics among the lay public, Dr. Feynman wrote *The Character of Physical Law* and *Q.E.D.: The Strange Theory of Light and Matter.* He also authored a number of advanced publications that have become classic references and textbooks for researchers and students.

Richard Feynman was a constructive public man. His work on the Challenger commission is well known, especially his famous demonstration of the susceptibility of the O-rings to cold, an elegant experiment which required nothing more than a glass of ice water. Less well known were Dr. Feynman's efforts on the California State Curriculum Committee in the 1960s where he protested the mediocrity of textbooks.

A recital of Richard Feynman's myriad scientific and educational accomplishments cannot adequately capture the essence of the man. As any reader of even his most technical publications knows, Feynman's lively and multi-sided personality shines through all his work. Besides being a physicist, he was at various times a repairer of radios, a picker of locks, an artist, a dancer, a bongo player, and even a decipherer of Mayan hieroglyphics. Perpetually curious about his world, he was an exemplary empiricist.

Richard Feynman died on February 15, 1988, in Los Angeles.